Illustrated Practical
Microbiology

Illustrated Practical
Microbiology

Amandeep Singh MBBS, MD
Postdoctoral Fellowship Resident
Christian Medical College, Vellore, TN
Ex-Assistant Professor
Department of Microbiology
Grant Government Medical College and JJ Group of Hospitals
Mumbai, Maharashtra

Shuvankar Mukhopadhyay MBBS, MD
Assistant Professor in Microbiology
Hitech Group of Medical Colleges, Odisha
Ex-Assistant Professor
Department of Microbiology
Dr Shankarrao Chavan Government Medical College
Nanded, Maharashtra

CBS

CBS Publishers & Distributors Pvt Ltd

New Delhi • Bengaluru • Chennai • Kochi • Kolkata • Mumbai
Hyderabad • Nagpur • Patna • Pune • Vijayawada

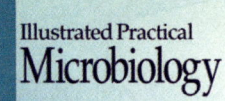

Illustrated Practical
Microbiology

ISBN: 978-93-86217-92-9

Copyright © Author and Publisher

First Edition: 2017

Published by Satish Kumar Jain and produced by Varun Jain for

CBS Publishers & Distributors Pvt Ltd

4819/XI Prahlad Street, 24 Ansari Road, Daryaganj, New Delhi 110 002, India.
Ph: 23289259, 23266861, 23266867 Website: www.cbspd.com
Fax: 011-23243014   e-mail: delhi@cbspd.com; cbspubs@airtelmail.in.
Corporate Office: 204 FIE, Industrial Area, Patparganj, Delhi 110 092
Ph: 4934 4934   Fax: 4934 4935   e-mail: publishing@cbspd.com; publicity@cbspd.com

Branches

- **Bengaluru:** Seema House 2975, 17th Cross, K.R. Road,
  Banasankari 2nd Stage, Bengaluru 560 070, Karnataka
  Ph: +91-80-26771678/79   Fax: +91-80-26771680   e-mail: bangalore@cbspd.com
- **Chennai:** 7, Subbaraya Street, Shenoy Nagar, Chennai 600 030, Tamil Nadu
  Ph: +91-44-26680620, 26681266   Fax: +91-44-42032115   e-mail: chennai@cbspd.com
- **Kochi:** Ashana House, No. 39/1904, AM Thomas Road, Valanjambalam,
  Ernakulam 682 018, Kochi, Kerala
  Ph: +91-484-4059061-65   Fax: +91-484-4059065   e-mail: kochi@cbspd.com
- **Kolkata:** 6/B, Ground Floor, Rameswar Shaw Road, Kolkata-700 014, West Bengal
  Ph: +91-33-22891126, 22891127, 22891128   e-mail: kolkata@cbspd.com
- **Mumbai:** 83-C, Dr E Moses Road, Worli, Mumbai-400018, Maharashtra
  Ph: +91-22-24902340/41   Fax: +91-22-24902342   e-mail: mumbai@cbspd.com

Representatives

- **Hyderabad** 0-9885175004 • **Nagpur** 0-9021734563 • **Patna** 0-9334159340
- **Pune** 0-9623451994 • **Vijayawada** 0-9000660880

Printed at International Print-o-Pac Limited, Noida, India

to

*our Teachers, Parents and Students
with Special Thanks to Dr Chhaya Chande*

"Never regard study as a duty, but as the enviable opportunity to learn."

Albert Einstein

Foreword

I take great pleasure in writing this foreword for *"Illustrated Practical Microbiology"*.

I know the authors from the last 4 years. Both are ambitious, highly motivated students and always eager to learn new things. I have observed in them a constant desire to learn more and strive for perfection. They have an experience in teaching microbiology to undergraduates and other paraclinical branches. Authors are also well versed with conducting undergraduate examinations for MBBS students.

The book includes many coloured photographs, microscopic slides, media and biochemical reactions, etc.

The book will be of definite benefit to students of MBBS, other paraclinical branches as well as postgraduates preparing for the university examinations.

I am delighted to encourage and assist them in pursuit of their goal. I wish the authors a grand success.

Abhay Chowdhary
MD, DHA, DM Virology, FIMSA, FRSTMH
Professor and Head
Department of Microbiology
Grant Govt. Medical College and Sir JJ Group of Hospitals
Mumbai

Preface

In our book titled *"Illustrated Practical Microbiology"*, the primary goal is to provide the undergraduate students with a reliable, comprehensive, practical and informative approach towards their practical examination. The authors describe the basic issues in practical microbiology, general principles for safety and diagnostic approaches including antimicrobial susceptibility testing and identification.

To facilitate understanding of the subject matter, we continue to employ the tables, detailed charts, summary text boxes and mnemonics. The labelled and captioned photographs have been selected to aid retention by engaging the visual memory in a manner complementary to mnemonics. The book is concise, easy to read, to the point text and a separate section on frequently asked questions has been added which has made the understanding of the subject easy and interesting. Also there is a relevant addition in the view of growing importance to hospital infection control measures and biomedical waste management.

This edition may replace the other books as a textbook of practical microbiology for MBBS, lab technology students and will also help the postgraduate students and thus may be recommended for the same. This edition integrates the theoretical principles and experimental techniques common to all the undergraduate courses.

We invite the students and faculty to share their thoughts and ideas to help us continually to improve through our blog and collaborative editorial platform. We will gladly make corrections if they are brought to our attention.

Amandeep Singh

Shuvankar Mukhopadhyay

Acknowledgements

All praise and glory to Almighty who gave us courage and patience to carry out this work. We would like to express our gratitude to all those who gave us the possibility to complete this book.

First of all, we would like to express our thanks to Dr Chhaya Chande (Associate Professor, GGMC, Mumbai) for her suggestions and provision of the materials evaluated in this book, whose expertise, understanding and patience, added considerably to our experience. We appreciate her vast knowledge and skill, and her assistance in writing this book.

We are very grateful to Dr SG Joshi (Professor, GGMC, Mumbai) and Dr Sarala Menon (Associate Professor, GGMC, Mumbai), without whose motivation and encouragement we would not have considered this book to be complete.

Dr Abhay Chowdhary (Professor and Head, Department of Microbiology, GGMC, Mumbai) is the one teacher who truly made a difference in our lives. It was under his tutelage that we developed a focus and became interested in writing this book. It was through his persistence, understanding and kindness that we completed this book. We doubt that we will ever be able to convey our appreciation fully, but we owe him our eternal gratitude.

Dr Jayanti Shastri (Professor and Head, Department of Microbiology, Nair Medical College, Mumbai) whose untiring efforts and drive to reach perfection have always been a source of inspiration. Her word of care and hand of support has been the pillar of strength while writing this book.

We would also like to thank our friends in the department for valuable, exchanges of knowledge while writing this book, which helped enrich the experience.

We would also like to thank our family with Special thanks to Dr Tapan Mukhopadhyay and Dr Ramala Mukhopadhyay, parents of second author and Mr Harvinder Singh and Ms Surinder Kaur, parents of first author for their support. Without their love, patience, understanding and encouragement we would not have been able to complete this book.

Department of Microbiology, Grant Govt Medical College and JJ Group of Hospitals, Mumbai; we will like to acknowledge following for providing us some of the illustrations:

Department of Microbiology, TNMC and BYL Nair Hospital, Mumbai.

Gupta Polyclinic and Diagnostic Centre, Rourkela, Odisha.

Last but not the least we would like to express our gratitude to CBS Publishers and their team for their substantial contributions associated with the completion of this book.

We hope this book serves to repay their contributions in some small part.

Amandeep Singh
Shuvankar Mukhopadhyay

Contents

Chapter 1

Microscopy

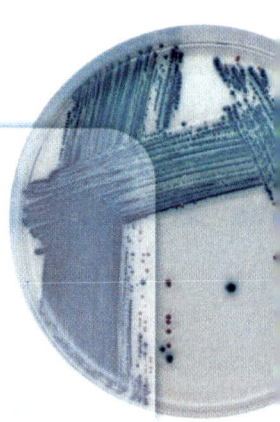

INTRODUCTION

Microorganisms are extremely small. The size of the bacteria is expressed in micrometer whereas viruses are measured in nanometer.

- Unit of measurement in microbiology:
  - 1 micron = 10^{-3} mm
  - 1 nanometer = 10^{-3} micron.
- Most of the bacteria of medical importance generally measure 0.2–1.5 micron in diameter and 3–5 micron in length contrast from the majority of human pathogenic viruses which range from 20–300 nanometers.
- Because of the small size, microorganism cannot be seen distinctly with the unaided eye (unaided eye can see not less than 0.2 mm smaller object). If we want to see object smaller than 0.2 mm, we should have to use microscope.
- Most of the bacteria can be observed by light microscope but viruses need an electron microscope.

TYPES OF MICROSCOPE

1. Bright field or light microscope
2. Dark field microscope
3. Phase contrast microscope
4. Fluorescence microscope
5. Electron microscope

Bright Field or Light Microscope

The bright field or light microscope forms a dark image against brighter background.

First microscope invented by Anton van Leeuwenhoek containing a single lens (Fig. 1.1) in 1600 BC.

Leeuwenhoek
microscope
(circa late 1600s)

Fig. 1.1: Leeuwenhoek microscope

Parts of Light Microscope

- Supporting system
- Adjustment system
- Illumination system
- Optical system

Supporting system

1. Base
2. Limb (arm)
3. *Mechanical stage*: It is fitted with fine vernier graduations as on a ruler.

Adjustment system

1. *Coarse and fine adjustment screw*: One revolution of the fine focusing knob should generally move the mechanical stage by 100 micron
2. Condenser adjustment and centering screw
3. Mechanical stage control
4. Iris diaphragm lever

Illumination system
1. *Mirror*—has two surfaces, one plain for oil immersion and other concave for dry lens (if source of light natural)
2. *Source of light*—natural/artificial
3. *Filters*—blue/green (blue is commonly used).

Optical system
1. *Objective lens*:
   • Objectives with magnifying powers 4×, 10×, 40× and 100× are commonly used. The 100× objective is for oil immersion.
   • The magnifying power is marked on the lens and is usually color-coded for easy identification.
   • Color coding of objective lens
     – Yellow—low power
     – Blue—high power
     – Black—oil immersion
2. *Eyepieces*:
   • Magnify real image
   • Carry scales and marker
   • Two planoconvex lenses with a circular field diaphragm
   • *Types*:
     – Huygenian (most commonly used)
     – Ramsden (for micrometry)

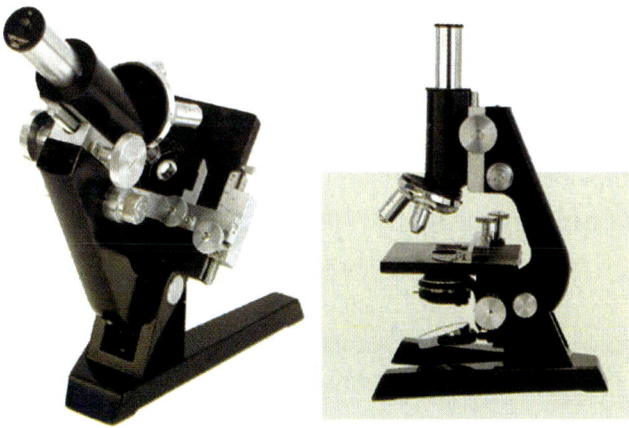

Fig. 1.2: Monocular microscope

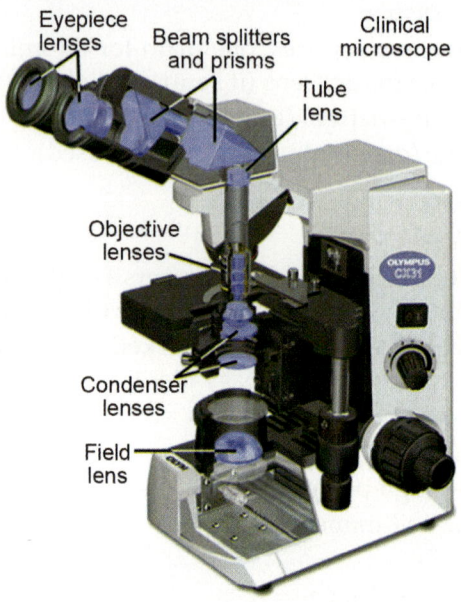

Eyepiece lenses

Beam splitters and prisms

Clinical microscope

Tube lens

Objective lenses

Condenser lenses

Field lens

Fig. 1.3: Binocular microscope

3. *Condenser* (Fig. 1.4):
- The condenser illuminates the specimen and controls the amount of light and contrast.
  - Numerical aperture (NA) is the light gathering power of the lens. NA of a condenser should be equal to or greater than that of the objective with maximum NA.
- An iris diaphragm is provided below the condenser, which adjusts the NA of the condenser when using objectives having low magnifying power.
- *Type of condenser*
  - Abbe condenser: Most commonly used.
  - Achromat condenser.
- *Cone of illumination.* The substage condenser must be focused and the diaphragm adjusted so that the cone of illumination completely fills the aperture of the microscope objective.

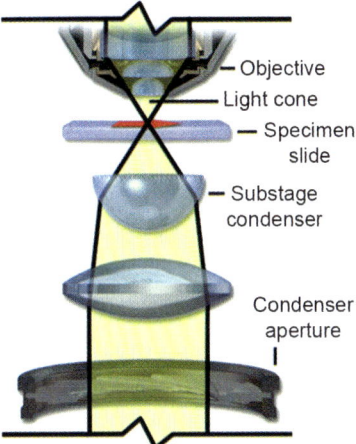

Fig. 1.4: Function of condenser: Focusing light ray on to sample object

- **Principle/mechanism of image formation** (Fig. 1.5). In the compound microscope, the intermediate image formed by the objective and tube lens is enlarged by the eyepiece.

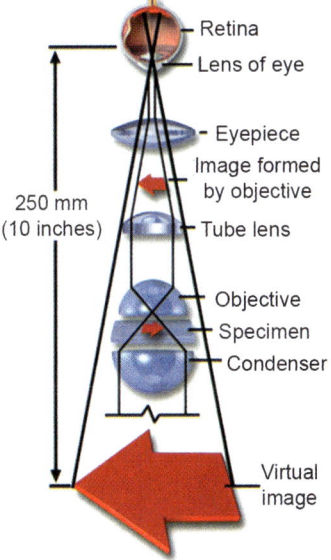

Fig. 1.5: Optics of image formation in light microscope

Magnification

- Magnification = Size of image/Size of object
- Magnification = Distance of image from objective/Distance of object from objective
- Magnification = Optical tube length/Focal length of objective
  - For low power – 160/16 = 10×
  - For high power – 160/4 = 40×
  - For oil immersion – 160/2 = 80×
- Total magnification = Objective × Eyepiece
  - *Numerical aperture*:
  - NA = Diameter of lens/Focal length
  - Expressed as 'NA = n Sin U',
    Where: n = Refractive index of medium (RI)
    (2U = angle of aperture)
  - Theoretical limit of U is 90°
  - RI of air is 1
  - NA of dry lens 1 × Sin 90 = 1 × 1 = 1

  The numerical aperture of the entire MICROSCOPE SYSTEM depends on the NA of the substage condenser and the objective working together.
- *Numerical aperture and light gathering ability* (Fig. 1.6)

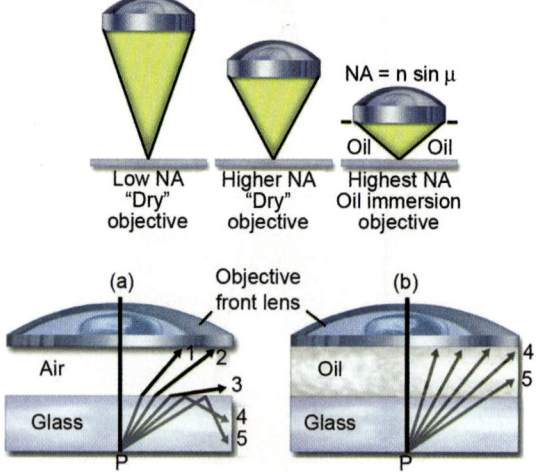

Fig. 1.6: Numerical aperture of different lens and advantage of oil immersion lens: More light gathering on to objective lens

- *Principle of oil immersion*: Cedar wood oil or previously used Canada balsam oil has same refractive index as that of the glass of the slide thus more light gathers on to objective lens which leads to more numerical aperture, hence more resolution.

Difference between Resolution and Definition of Microscope

Resolution power of the lens
- Ability to reveal closely adjacent structural details as separate and distinct
- Limit of useful magnifying power
- Depends on the wavelength of light and NA of the objective
- **Resolution** of lens = 0.61 × wavelength/NA
  For green light – 0.61 × 0.55/1.4 = **0.24 µm** (240 nm)
- Thus resolution of light microscope is 0.2 micron
- Highest resolution with oil immersion. (NA of oil immersion >NA of high power >NA of low power).

Definition of microscope
- Defined as capacity of an objective to render the outline of the image of an object clear and distinct.
- Spherical aberrations occur because of uneven thickness of lenses.
- Chromatic aberrations are caused due to splitting of white light.
  - These aberrations can be corrected by the use of crown glass, white flint glass and achromatic lenses.

Care of the Microscope

Do's
- Keep at uniform temperature
- Keep it covered
- Clean it frequently
  - Soft camel hair brush, fine tissue paper, muslin silk/cotton cloth
  - Benzol/xylol for oil immersion.

Don'ts
- Use gauze piece and cotton for cleaning lenses
- Expose to direct sunlight
- Use alcohol or acetone for cleaning lenses

Maintenance

It is to be done only by the authorized personnel.

Uses

- Scanner (4×)—to view a large area at a glance.
- Low and high power—unstained preparations (wet mount), negative stains, peripheral smears.
- Oil immersion—stained preparations to study morphology of microorganisms.

Application of Microscopy

- One of the direct methods of diagnosis.
- Viewing of unstained and stained preparations.
- Of vital importance in certain emergencies such as cholera, diphtheria, gas gangrene, etc.
- Gives a clue of the etiologic agent and aids in follow-up.
- Acid-fast bacilli (AFB) staining is one of the cornerstones in tuberculosis (TB) diagnosis in rural areas.

Inverted Microscope (Fig. 1.7)

Differences from the optical microscope are:
- Here the light sources and the condenser are on the top of the stage.
- The eyepiece is not upside but places standard viewing angle.
- *Use*:
  - To visualize tissue culture and cytopathic effect in virology.
  - To observe hemolysis in bacteriology and MODS assay in TB.

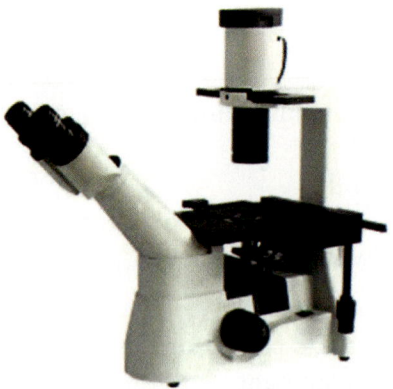

Fig. 1.7: Inverted microscope

Dark Ground Microscope

- In optical microscopy, dark field describes an illumination technique **used to enhance the contrast in unstained samples**. It works by illuminating the sample with light that will not be collected by the objective lens, and thus will not form part of the image. This produces the classic appearance of a dark, almost black, background with bright objects on it.
- Fig. 1.8 illustrating the light path through a dark field microscope.
- Only the **reflected light** goes on to produce the image.
- *Characteristics of a dark ground microscope*:
  - Dark ground condenser containing patch stop.
  - High intensity lamp.
- *Application*: To identify living unstained cells and very thin bacteria like spirochetes which cannot be visualized by light microscopy.

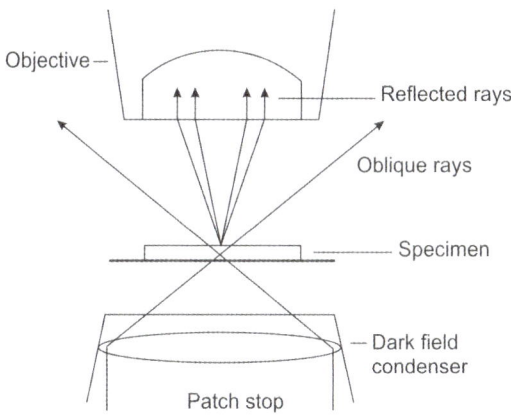

Fig. 1.8: Light pathway of dark ground microscope

Phase Contrast Microscope

The phase contrast microscope is widely used for examining specimens such as biological tissues. It is a type of light microscopy that **enhances contrasts of transparent and colorless objects by influencing the optical path of light**. The phase contrast microscope is able to show components in a cell or bacterium, which would be very difficult to see in an ordinary light microscope.

Altering the Light Waves

The phase contrast microscope uses the fact that the light passing through a transparent part of the specimen travels slower and, due to this, is shifted compared to the uninfluenced light. This **difference in phase** is not visible to the human eye. However, the change in phase can be increased to half a wavelength by a transparent **phase-plate** in the microscope and thereby causing a difference in brightness. This makes the transparent object shine out in contrast to its surroundings (Fig. 1.9).

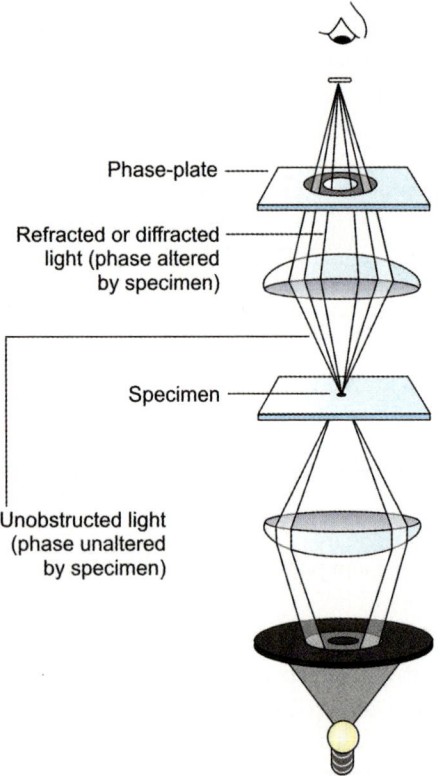

Phase-plate

Refracted or diffracted light (phase altered by specimen)

Specimen

Unobstructed light (phase unaltered by specimen)

Fig. 1.9: Light pathway of phase contrast microscope

Application

- The phase contrast microscope has made it possible to **study living cells and cell division** is an example of a process that has been examined in detail with it.
- Detect **microbial motility**.
- Detect **bacterial component like inclusion body and endosome**.

Fluorescence Microscope (Fig. 1.10)

The specimen is illuminated with light of a specific wavelength (usually **ultraviolet rays**) which is absorbed by the **fluorophores**, causing them to emit light of longer wavelengths (i.e. of a different color than the absorbed light). The illumination light is separated from the much weaker emitted fluorescence through the use of a spectral emission filter. Typical components of a fluorescence microscope are:

- A light source **(xenon arc lamp or mercury-vapor lamp)** is common; more advanced forms are high-power **light emitting diodes (LED)** and lasers.

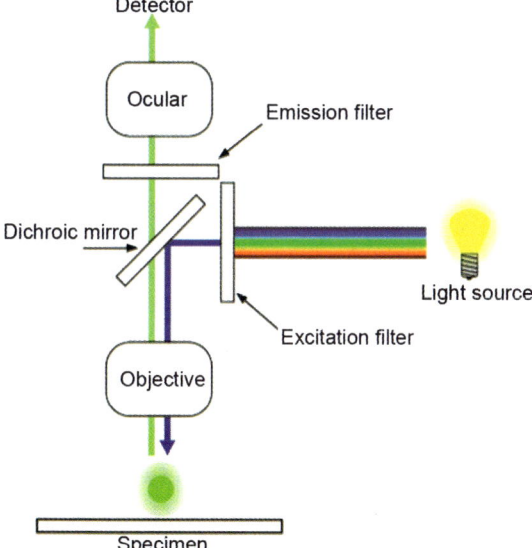

Fig. 1.10: Light pathway of fluorescence microscope

- The excitation filter.
- The **dichroic mirror (dichroic beam splitter).**
- Emission filter.

Application

- **Auramine phenol** is used for detection of **tubercle bacilli** which take apple green color.
- **Acridine orange** dye used for detection of **malaria parasite in quantitative buffy coat (QBC).**
- **In direct and indirect immunofluorescence tests, fluorescent dye tagged immunoglobulin** is used.
- **Other examples of fluorescent dyes**—fluoroscein iso-thiocyanate (FITC) and rhodamine.
- **Picture showing**—*Mycobacterium tuberculosis* stained with aura-mine dye and seen under fluorescence microscope (Fig. 1.11).

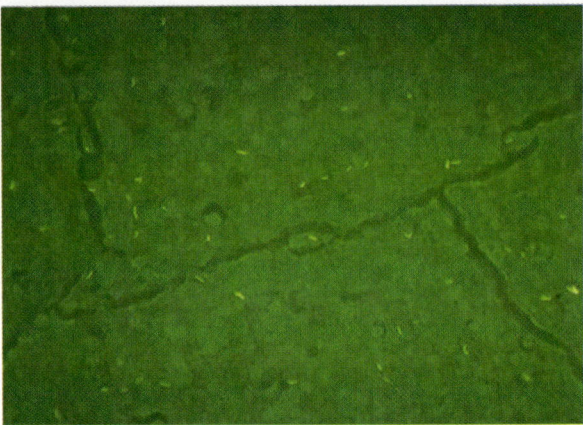

Fig. 1.11: *M. tuberculosis* under fluorescence microscope

Electron Microscope

- Earnst Ruska with his mentor Max Knoll built the first electron microscope in 1931.
- When it is required to **visualize viruses** or parts of cell diameter less than 0.2 μm it is necessary to use the EM.
- Limit of resolution 1 nm or less.
- Object is scanned with high-speed **electron.**

- Column is evacuated to prevent scattering of electron by air gas particle.
- Electron beam is **focused with electromagnetic lenses.**
- Wavelength of electron beam 0.005 nm or 100000 times shorter than visible light. Thus theoretically the **limit of resolution should be 100000 times greater than optical microscope but practically it is less** "due to narrow diameter of electron microscope tube."

  Two types:
  - Transmission electron microscope
  - Scanning electron microscope.
- *Transmission EM*—electron is transmitted **through specimen** material which **scatters electron** and produces image.
- *Scanning EM*—image is formed from **electron emitted by material surface** so **surface structure** is better visualized.

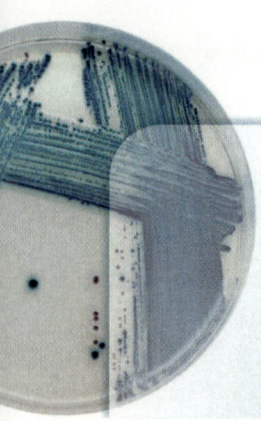

Chapter 2

Staining Methods

Stain: A stain is a chemical compound used to enhance the visibility of a microscopic object or organism.

TYPES OF STAINING TECHNIQUES

Simple Staining Technique

Principle: In a simple staining technique, a basic, cationic dye is flooded across a sample, adding color to the cells. A cation is simply a positively charged ion. The molecules that make up basic dyes have a positive charge. This is important because the cell wall and cytoplasm of bacterial cells have a negative charge. The positively charged dye is attracted to the negatively charged cells, enhancing the ability of the stain to stick to and color the cells. *Example*: Methylene blue.

Negative Staining Technique

Principle: In a negative staining technique, an acidic, anionic dye is mixed with a cell sample. The dye changes the color of the background, not the cells, causing the cells to stand out. This process can be considered the opposite of simple staining.

An anion is a negatively charged ion, therefore, an anionic dye has a negative charge. When the negatively charged dye is added to the negatively charged cells, the two repel each other, meaning they push apart. When the mixture is placed on a slide and air dried, what results is a darkly dyed background, surrounding clear, unstained cells. *Example*: India ink—used for demonstration of capsule of *Cryptococcus* and capsule of *Streptococcus pneumoniae*.

Differential Staining Technique

Principle: A procedure that allows the observer to visually distinguish between different types of bacterial cells based on the idea that not all cell types stain equally. This technique takes advantage of the different physical properties that different bacteria have evolved. *Examples*:

• Gram stain
• Acid-fast stain
• Albert stain

Gram Staining

Gram staining method, the most important procedure in microbiology, was developed by Danish physician Hans Christian Gram in 1884. Gram staining is still the cornerstone of bacterial identification and taxonomic division.

Principle of Gram stain: The differences in cell wall composition of Gram-positive and Gram-negative bacteria account for the Gram staining differences. Gram-positive cell wall contains thick layer of peptidoglycan with numerous teichoic acid cross linking which resists the decolorization.

This differential staining procedure separates most bacteria into two groups on the basis of cell wall composition:

1. Gram-positive bacteria (thick layer of peptidoglycan-90% of cell wall)—stain purple.
2. Gram-negative bacteria (thin layer of peptidoglycan-10% of cell wall and high lipid content)—stain pink.

Nearly all clinically important bacteria can be detected using this method the only exceptions being those organisms:

1. That exist almost exclusively within host cells, i.e. intra-cellular bacteria (e.g. *Chlamydia spp*).
2. That lack a cell wall (e.g. *Mycoplasma spp*).
3. Those of insufficient dimensions to be resolved by light microscopy (e.g. *Spirochaetes spp*).

Gram staining technique (Hucker's modification) involves following steps (Fig. 2.1):

1. Fixation of clinical materials to the surface of the micro-scope slide either by heating.

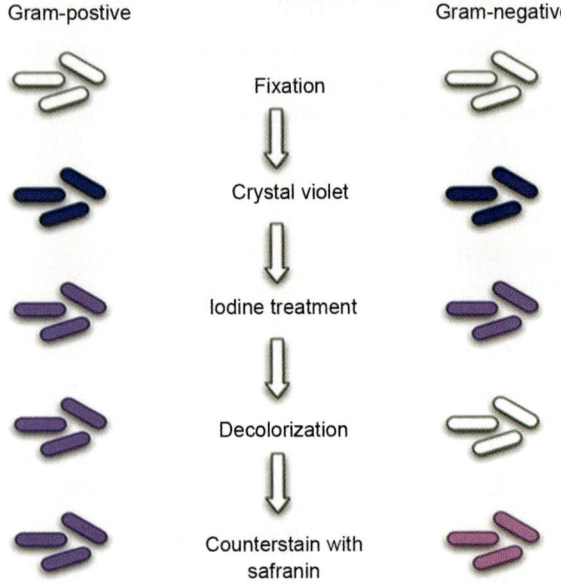

Fig. 2.1: Steps of Gram stain

2. Application of the **primary stain (crystal violet)** for 1 minute. Crystal violet stains all cells blue/purple.

3. Application of **mordant**: The **iodine solution** (mordant) is added for 1 minute to form a crystal violet iodine (CV-I) complex; all cells continue to appear blue.

4. **Decolorization step:** In the decolorization step, **acetone** is added dropwise for **2 to 3 seconds**. This step distinguishes Gram-positive from Gram-negative cells. The organic solvent, such as acetone or ethanol, extracts the blue dye complex from the lipid-rich, thin-walled Gram-negative bacteria to a greater degree than from the lipid poor, thick-walled, Gram-positive bacteria. The Gram-negative bacteria appear colorless and Gram-positive bacteria remain blue.

5. Application of **counterstain (safranin)** for 30 seconds: The red dye safranin stains the decolorized Gram-negative cells red/pink; the Gram-positive bacteria remain blue.

FAQs

• *Modifications of Gram stain*:

Stain	Primary stain	Decolorizer	Counterstain
Kopeloff and Beerman modification	Methyl violet	Acetone alcohol	Basic fuchsin
Jensen modification (for *Neisseria* spp)	Methyl violet	Absolute alcohol	Neutral red
Weigert modification (for tissue section)	Carbol gentian violet	Aniline xylol	Carmalum solution
Preston Morrel modification	Crystal violet	Iodine-acetone	Carbol fuchsin

• *Other primary stains used*:
  – Methyl violet
  – Gentian violet
  – Aniline gentian violet
• *Other decolorizers used*:
  – Acetone alcohol
  – Absolute-alcohol
• *Other counterstains used*:
  – Neutral red
  – Picric acid
  – Malachite green

Acid-Fast Stain

• Acid-fast stain is the differential staining technique which was first developed by Ziehl and later on modified by Neelsen. So this method is also called Ziehl-Neelsen staining technique. The main aim of this staining is to differentiate bacteria into acid-fast group and non-acid-fast group.
• This method is used particularly for the member of genus *Mycobacterium* (Fig. 2.2).

Principle of acid-fast stain

• Acid-fast cells resist decolorization with H_2SO_4 due to the presence of large amount of lipoidal material (mycolic acid) in their cell wall which prevents the penetration of decolorizing solution.

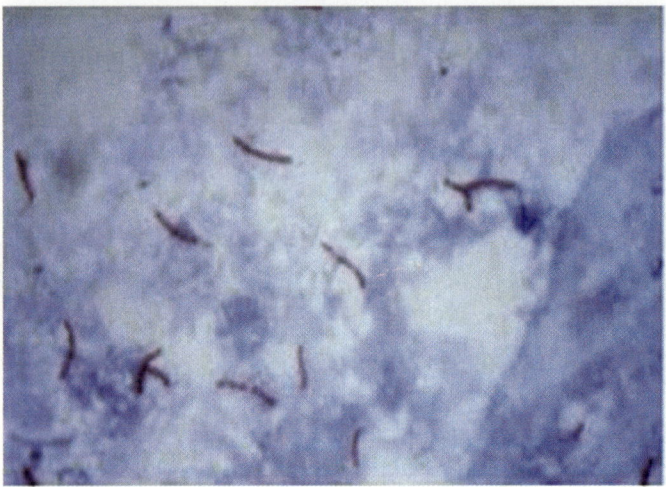

Fig. 2.2: *M. tuberculosis* stained with acid-fast stain and seen under oil immersion

- The non-acid-fast organisms lack the lipoidal material in their cell wall due to which they are easily decolorized, leaving the cells colorless. Then the smear is stained with counter-stain, methylene blue.
- Only decolorized cells absorb the counterstain and take its color and appear blue while acid-fast cells retain the red color.

Procedure of acid-fast stain
- *Hot method*:
  1. Prepare bacterial smear on clean and grease-free slide, using sterile technique.
  2. Allow smear to air dry and then heat fix.
  3. Cover the smear with carbol fuchsin stain.
  4. Heat the stain until vapor just begins to rise (i.e. about 60°C). Do not overheat. Allow the heated stain to remain on the slide for 5 minutes.
  5. Wash off the stain with clean water.
  6. Cover the smear with **25% sulphuric acid** for 5 minutes or until the smear is sufciently decolorized, i.e. pale pink.

7. Wash well with clean water.
8. Cover the smear with methylene blue stain for 1 minute, using the longer time when the smear is thin.
9. Wash off the stain with clean water.

Thus when the smear is stained with carbol fuchsin, it solubilizes the lipoidal material (mycolic acid) present in the *mycobacterial* cell wall but by the application of heat, carbol fuchsin further penetrates through lipoidal wall and enters into cytoplasm. Acid-fast cells resist decolorization with H_2SO_4, thus after all cells appear red.

- *Cold method (Kinyoun stain)*: Every step is similar to hot method except slide is not heated after pouring carbol fuschin.

Acid-fast organism	Percentage of sulphuric acid used as decolorizer
Mycobacterium tuberculosis	25% (as per recent RNTCP guideline)
Mycobacterium leprae	5%
Cryptosporidium, Isospora and Cyclospora, Microsporidium	10%
Nocardia	1%
Taenia saginata segment and eggs	1%
Hooklet of hydatid cyst	1%
Eggs of Schistosoma mansoni	1%
Bacterial spore	0.25–0.5%
Sperm head	0.5–1%
Legionella micdadei	0.5–1%

Albert Stain

- A differential stain used for staining the volutin granules also known as metachromatic granules or food granules found in *Corynebacterium diphtheriae*. It is named as meta-chromatic because of its property of changing color, i.e. when stained with blue stain it appears red in color.
- When grown in Loffler's slopes, *C. diphtheriae* produces large number of granules (Fig. 2.3).

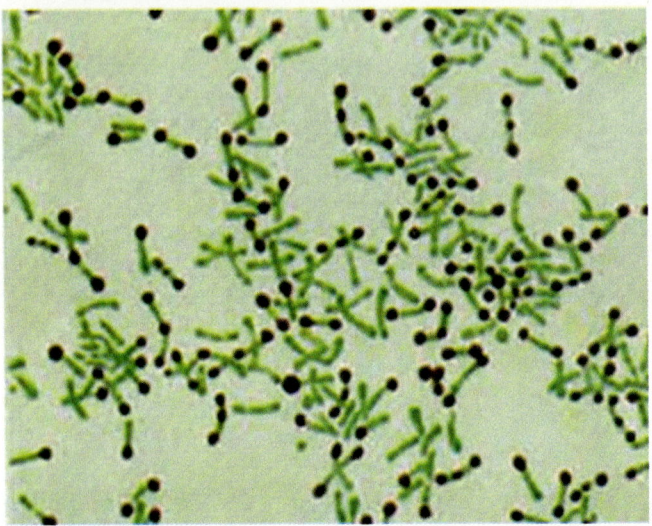

Fig. 2.3: *Corynebacterium diphtheriae* stained by Albert stain

Principle of Albert staining

1. Albert stain is basically made up of two stains that are toludine blue 'O' and malachite green, both of which are basic dyes with high affinity for acidic tissue components like cytoplasm. The pH of Albert stain is adjusted to 2.8 by using acetic acid which becomes basic for volutin granules as pH of volutin granule is highly acidic.

2. Therefore, on applying Albert's stain to the smear, toludine blue 'O' stains volutin granules, i.e. the most acidic part of cell and malachite green stains the cytoplasm blue-green.

3. On adding Albert's iodine, due to effect of iodine, the metachromatic property is not observed and granules appear blue in color.

Composition of Albert's stain

1. *Albert's A solution consists of:*
   - Toludine blue   0.15 gm
   - Malachite green   0.20 gm
   - Glacial acetic acid   1 ml
   - Alcohol (95% ethanol)   2 ml

2. *Albert's B solution consists of:*
   - Iodine   2 gm
   - Potassium iodide (KI)   3 gm

Procedure

1. Prepare a smear on clean grease-free slide. Air dry and heat fix the smear.
2. Treat the smear with Albert's stain and allow it to react for about 7 minutes.
3. Drain off the excess stain, do not wash the slide with water.
4. Flood the smear with Albert's iodine for 2 minutes.
5. Wash the slide with water, air dry and observe under oil immersion lens.

Result: If *Corynebacterium diphtheriae* is present in the sample, it appears green-colored, rod-shaped bacteria arranged at angle to each other, resembling English letter 'L', 'V' or Chinese letter pattern along with bluish black metachromatic granules at the poles.

Uses: This helps to distinguish *Corynebacterium diphtheriae* from most of the short nonpathogenic diphtheroides which lack granules.

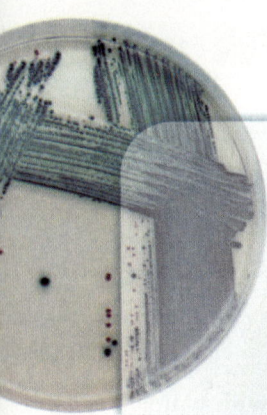

Chapter 3

Slides

The university practical examination for undergraduates also includes slide identification. These are some of the most commonly asked slides. All the pictures are original. Each slide includes some frequently asked questions which will mainly help students in their viva, so that they can score better.

GRAM-POSITIVE COCCI (GPC)

Different arrangements of cocci are as follows:
1. GPC in clusters (Fig. 3.1): *Staphylococci*
2. GPC in pairs and chains (Fig. 3.3): *Streptococci*
3. GPC in pairs: *Enterococci, pneumococci*
4. GPC in tetrads (Fig. 3.2): *Micrococci*
5. GPC in octads: *Sarcina*

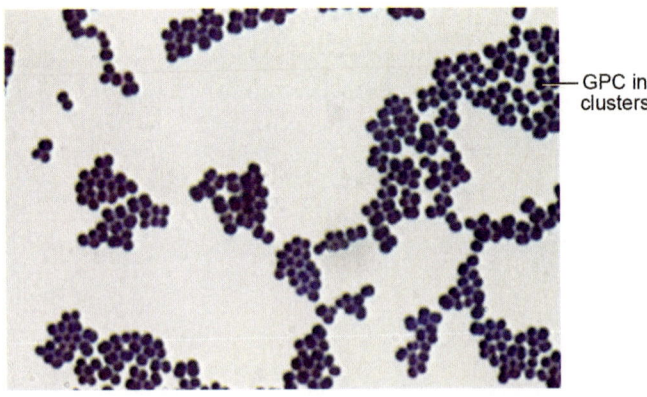

GPC in clusters

Fig. 3.1: Gram-positive cocci in cluster

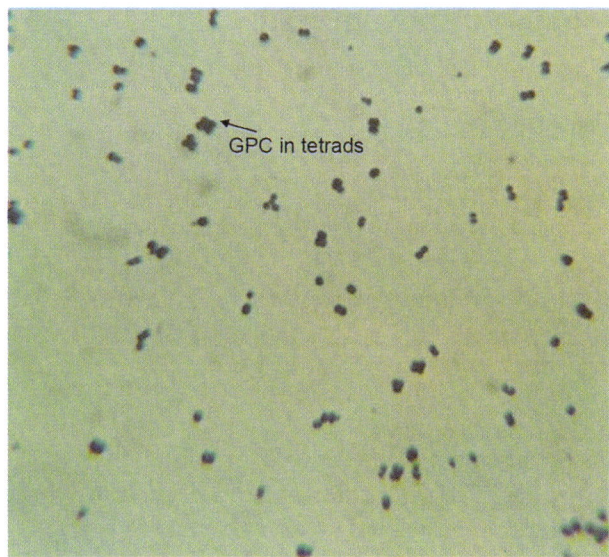

Fig. 3.2: Gram-positive cocci in tetrad

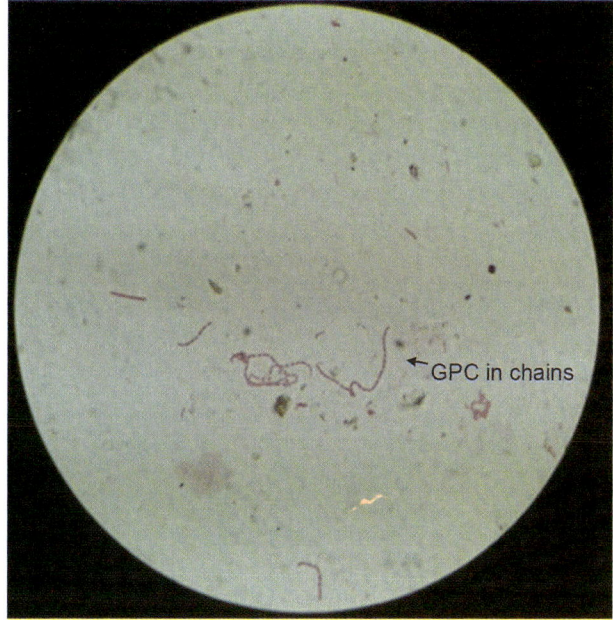

Fig. 3.3: Gram-positive cocci in chain

Staphylococcus

GPC in clusters, catalase positive, size—1 μm.

Diseases caused

1. *Skin and soft tissue*: Folliculitis, furuncle, carbuncle, abscess, wound infection, impetigo, less often cellulitis.
2. *Musculoskeletal*: Osteomyelitis, arthritis, bursitis, pyomyositis.
3. Urinary tract infection (UTI).
4. Toxic shock syndrome, staphylococcal scalded skin syndrome.
5. Food poisoning.
6. Sepsis.

Media

1. Hemolysis shown—β-hemolysis on blood agar
2. Selective medium—Ludlam's medium (lithium chloride and tellurite), mannitol salt agar (7.5% NaCl).
3. Potassium tellurite medium—growth on this shows black-colored colonies.

FAQs

Various species of *Staphylococcus* and their characteristic features:

1. *Staphylococcus aureus*: Golden-yellow pigment, both slide and tube coagulase positive, shows mannitol fermentation
2. *S. epidermidis*: Slide and tube coagulase negative
3. *S. saprophyticus*: Causes UTI in sexually active females

Staphylococcus species producing pigments:
1. *S. aureus*—golden-yellow

Staphylococcus species tube coagulase positive:
1. *S. aureus*
2. *S. intermedius*
3. *S. hycius*

Staphylococcus species slide coagulase positive but tube coagulase negative:
1. *S. lugdunensis*
2. *S. schleiferi* sub-sp *coagulans*

Antimicrobial susceptibility testing:

1. Penicillin is drug of choice (DOC) for penicillin-sensitive strains of *Staphylococcus.*
2. Methicillin drug helps in identification of the methicillin-sensitive and methicillin-resistant *Staphylococcus.*
3. DOC for methicillin sensitive *Staphylococcus*: Nafcillin or oxacillin.
4. DOC for methicillin-resistant *Staphylococcus* (MRSA): Vancomycin.
5. Methicillin resistance is due to presence of **mec-A gene** which encodes for altered penicillin binding protein-2a **(PBP-2a).**
6. Gold standard test to diagnose MRSA—polymerase chain reaction (PCR).

Streptococci

GPC in chains, catalase negative, size—0.5–1 μm.

Diseases caused

1. *Viridians group streptococci*: Dental caries and subacute bacterial endocarditis (SABE).
2. *Pneumococcus*: Lobar pneumonia, meningitis, empyema and effusions.
3. Gp-*A Streptococcus pyogenes*: Pharyngitis, scarlet fever, impetigo, cellulitis, necrotizing fasciitis, bacteremia, toxic shock syndrome and two non-suppurative complications—acute rheumatic fever and glomerulonephritis.
4. Gp-B *Streptococcus agalactiae*: Puerperal sepsis and early and late onset neonatal meningitis.
5. *Enterococcus*: UTI, endocarditis, bacteremia.

Media

1. Blood agar
2. Chocolate agar
3. *Selective medium*: Crystal violet blood agar and PNF media (polymyxin, neomycin and fusidic acid).
4. *Transport medium*: Pike's medium.

FAQs

Classification:

1. *Lancefield classification:* Based on C carbohydrate antigen β-hemolytic streptococci are divided into 20 serogroups (A to V *except* I and J).

2. *Griffth classification:* Based on M-proteins β-hemolytic streptococci are divided into >80 serotypes.

Hemolysis shown by *Streptococcus* species:

1. α-hemolysis: *Viridians group of streptococci* (occur in chains) *Pneumococcus* (diplococci—show characteristic lanceolate shaped).

2. β-hemolysis: Gp-A *Streptococcus pyogenes*, Gp-B *Streptococcus agalactiae.*

3. γ-hemolysis: *Enterococcus* species.

   Examples of Fastidious organisms:

   1. *Streptococcus pneumoniae*
   2. *Neisseria* spp
   3. *Hemophilus influenzae*

Streptococci in liquid medium:

1. *Streptococcus pneumoniae*—uniform turbidity
2. *Viridans streptococci*—granular turbidity

Serological tests done to diagnose streptococcal infection:

1. Antistreptolysin O titres (ASO titres)—for diagnosis of acute rheumatic fever, **ASO titres >200 IU/ml** considered positive.

2. Anti-DNase test—for diagnosis of post-streptococcal glomerulonerphritis, **anti-DNase titres >300 IU/ml** considered positive.

3. CRP (C-reactive protein) is an abnormal protein that appears in the blood in the acute stages of various inflammatory disorders but is undetectable in the blood of healthy persons. The name of the protein was derived from the fact that it forms a precipitate with the non-type specific somatic C-polysaccharide of the *Pneumococcus*. It is also increased in sepsis by any bacteria and acute tissue injury (acute

myocardial infarction). It is most commonly used as a prognostic marker rather than diagnostic test. CRP levels of **0.6 mg/dl** considered as positive.

Special tests for *Streptococcus pneumoniae* detection:
1. *Bile solubility test*: Tube method using 10% sodium deoxycholate and slide method using 2% sodium deoxycholate is **positive**.
2. Optochin sensitive
3. Inulin fermentation positive
4. Quellung reaction (for capsule demonstration).

Special tests for *Enterococcus* detection:
1. Bile esculin positive
2. Grows in 6.5% NaCl
3. Grows in presence of 40% bile, at pH—9.6 and 45°C and 10°C
4. Heat tolerance test positive—survive 60°C for 30 minutes.

Antimicrobial susceptibility testing:
1. DOC for *streptococci* infection—Penicillin
2. DOC for severe infection like, endocarditis by *enterococci*, is Penicillin + Aminoglycosides combination therapy and if *enterococci* is resistant to penicillin then DOC vancomycin.
3. For vancomycin-resistant *enterococci* (VRE), drug given is daptomycin.

GRAM-NEGATIVE COCCI (GNC)

Neisseria Meningitidis (Fig. 3.4)

Capsulated Gram-negative, **lens-shaped diplococci** (0.5–1 μm), catalase positive, oxidase positive.

Diseases caused
1. Meningitis
2. Meningococcal septicemia
3. Fulminant meningococcemia—Waterhouse-Friderichsen syndrome characterized by purpuric rashes, disseminated intravascular coagulation (DIC), bilateral adrenal hemorrhage and multiorgan failure.

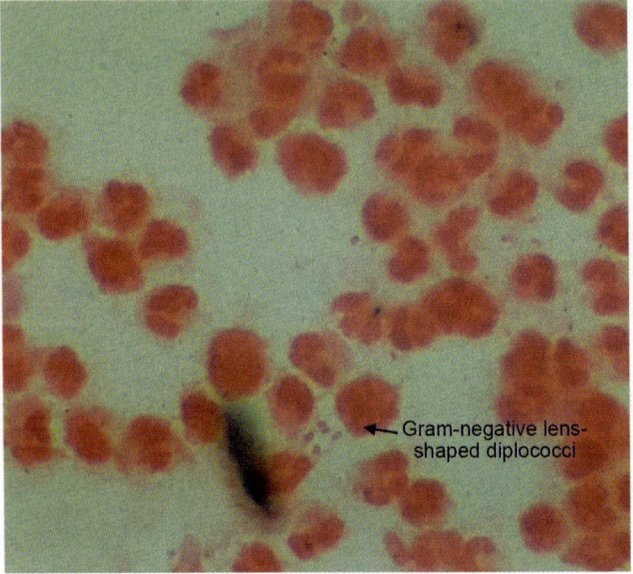

Gram-negative lens-shaped diplococci

Fig. 3.4: Gram-negative diplococci—lens shaped

Media
1. Blood agar
2. Chocolate agar
3. Selective medium: Modified Thayer-Martin medium, New York City medium.
   (**Note: Modified** Thayer-Martin medium contains antibiotics—vancomycin, colistin, nystatin, **trimethoprim.**)
4. Transport medium: Stuart's medium.

FAQs
1. Special tests for *N. meningitidis* detection: Rapid carbohydrate utilization test (RCUT)—ferment glucose and maltose.
2. *Treatment*:
   1. DOC: Ceftriaxone or cefotaxime
   2. Drugs for chemoprophylaxis: Rifampicin and ciprofloxacin
   3. Vaccine prophylaxis: Meningococcal **polysaccharide** vaccine—bivalent (serogroups A and C) or quadrivalent (serogroup A, C, Y, W-135) administered as 50 mg, 2 doses, 2–3 months apart to children aged 3–18 months.

Neisseria Gonorrhoeae (Fig. 3.5)

Non-capsulated, **Gram-negative, kidney-shaped diplococci** (0.5–1 µm), catalase positive, oxidase positive.

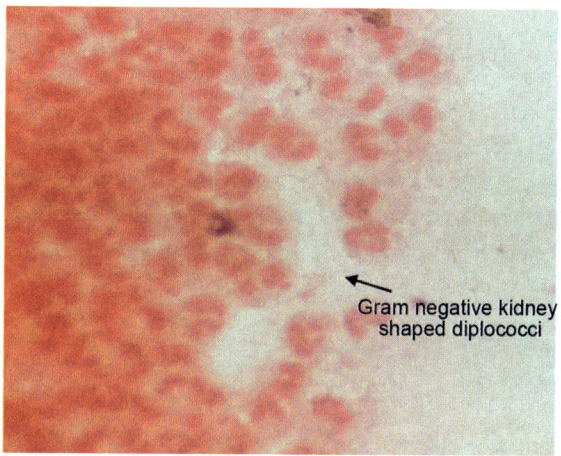

←Gram negative kidney shaped diplococci

Fig. 3.5: Gram-negative diplococci—kidney shaped

Diseases caused

1. *Sexually transmitted infection (STI)*: In males present as acute urethritis, epididymitis, prostatitis, etc. and in females present as mucopurulent cervicitis, **vulvovaginitis in prepubertal girls**, Fitz-Hugh-Curtis syndrome (peritonitis and perihepatic inflammation).

2. Ophthalmia neonatorum

3. *Disseminated gonococcal infection (DGI)*: Polyarthritis, tenosynovitis, pustular-hemorrhagic skin rash and rarely endocarditis.

Media

1. Blood agar

2. Chocolate agar

3. *Selective medium*: Modified Thayer-Martin medium, New York City medium, Martin-Lewis medium.

4. Transport medium—Stuart's medium and Amie's medium.

FAQs

1. Special tests for *N. gonorrhoeae* detection: Rapid carbohydrate utilization test (RCUT)—ferment only glucose.
2. *Treatment*:
   i. DOC—ceftrioxone or cefixime
   ii. No vaccine available for gonococci.
3. *Causes of non-gonococcal urethritis*: *Mycoplasma* , *Ureaplasma urealyticum*, *Chlamydia* sp, *Trichomonas vaginalis*.
4. Swabs used for gonococcal sample collection are calcium alginate swab, dacron, rayon, charcoal laden swabs but NOT cotton swabs because fatty acids present in cotton swab kill the gonococci organism.

GRAM-NEGATIVE BACILLI

Examples

1. *Escherichia coli*
2. *Klebsiella pneumoniae*
3. *Salmonella* spp
4. *Proteus* spp

Diseases caused

1. Skin and soft tissue abscesses, wound infections are caused by *E.coli, Klebsiella pneumoniae*.

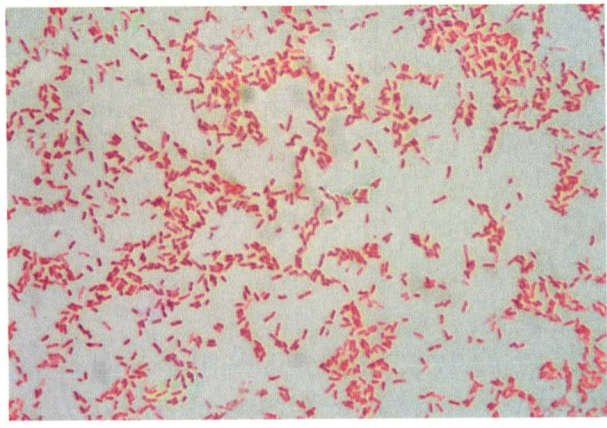

Fig. 3.6: Gram-negative bacilli

2. Urinary tract infection (UTI)—*E. coli, Proteus* spp, etc.
3. Food poisoning—*Salmonella* spp.

Media
1. Blood agar
2. MacConkey agar

FAQs
1. *Examples of motile bacteria*:
- *E. coli* due to peritrichous flagella
- *Salmonella* spp due to peritrichous flagella
- *Vibrio cholerae* due to polar flagella
- *Pseudomonas* spp due to polar flagella
- *Proteus* spp shows swarming

2. *Non-motile bacteria*:
- *Klebsiella* spp
- *Shigella* spp

3. *Motile parasites*:
- *Giardia lamblia*—falling leaf-like motility
- *Trichomonas vaginalis*—jerky motility

4. *Types of motility*:
- *Darting motility—Vibrio cholerae*
- *Tumbling motility—Listeria monocytogenes*

5. *Methods of detecting motility*:
- Hanging drop preparation
- Wet mount
- Semisolid agar—Craige's tube or U-tube
- Swarming
- Motility medium—SIM (sulphide indole motility medium), MIO medium (motility indole ornithine medium)
- Flagellar stain—Rye stain
- Silver impregnation staining
- Dark ground microscopy used for *Spirochetes* motility demonstration
- Electron microscopy

CORYNEBACTERIUM DIPHTHERIAE (KLEBS-LOEFFLER BACILLI)

Corynebacterium diphtheriae: Small, thin, slender, pleomorphic (club-shaped), **Gram-positive bacilli** and of size approximately 3–6 μm × 0.6–0.8 μm (Figs 3.7 and 3.8).

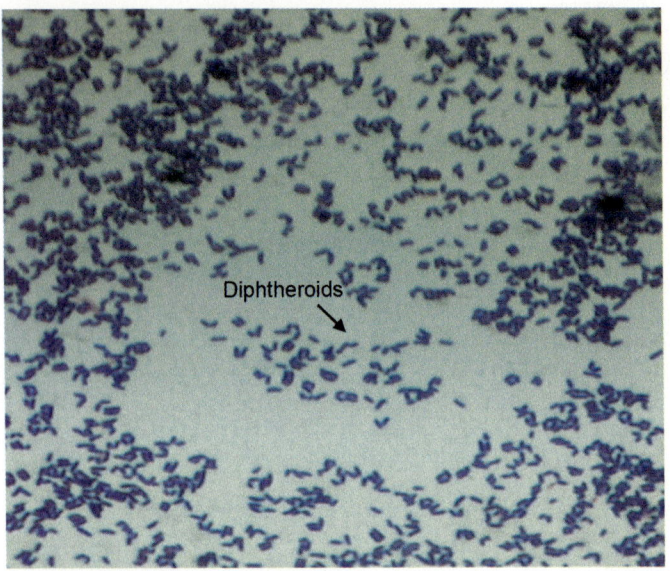

Fig. 3.7: Diphtheroid

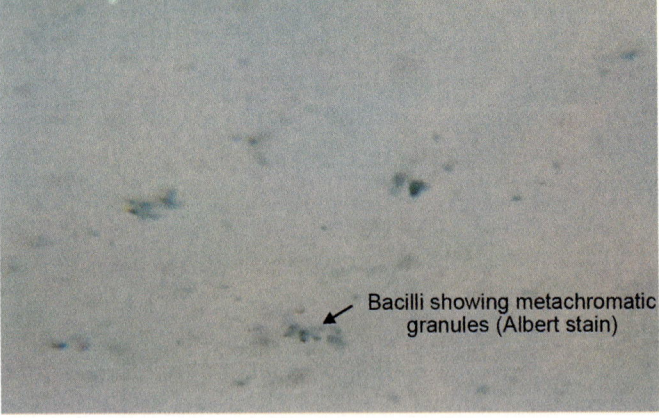

Fig. 3.8: *C. diphtheriae* seen by Albert stain

Special staining: **Albert's stain, Neisser's stain** and **Ponder's stain**—shows the presence of metachromatic granules also called as polar bodies or Babes-Ernst granules or volutin granules. These granules are seen at the ends of the characteristic Chinese letter or cuneiform arrangement of the bacilli.

Corynebacterium species showing granules: *C. diphtheriae* and *C. xerosis*.

Diseases caused
1. Respiratory diphtheria.
2. Cutaneous diphtheria
3. *Complications due to toxin*: Neurological and myocarditis.

Media
1. Blood agar
2. Loeffler's serum slope (growth appears in 6–8 hrs, best medium for demonstration of metachromatic granules)
3. Selective medium—potassium tellurite medium [(blood agar + potassium tellurite **(0.04%)**] and Tinsdale medium (potassium tellurite + cysteine)

FAQs
1. *Diphtheria toxin*: β-corynephage encoded carrying tox gene and acts by inhibiting protein synthesis.
2. *Treatment*:
   i. Antidiphtheritic serum: IM or IV; dose 20,000–40,000 units for mild pharyngeal infection to 1 lakh units in severe cases.
   ii. *DOC*: Penicillin or erythromycin. Most useful if administered within 6 hrs of infection, i.e. before toxin release.
   iii. *DOC for carriers*: Erythromycin.

GRAM-POSITIVE BACILLI

Examples
1. *Bacillus* species
2. *Clostridium* species

Bacillus Species

Gram staining: Chains of Gram-positive, **large** bacilli with **non-bulging spores** (Fig. 3.9).

Bacillus species showing lipid granules: B. anthracis which is demonstrated by Sudan black-B.

Diseases caused

1. Anthrax: *Bacillus anthracis.*
2. Food poisoning manifested as episodes of vomiting within 5 hrs of Chinese fried rice consumption: *Bacillus cereus (emetic type).*
3. Food poisoning manifested as episodes of diarrhea after 12–24 hrs: *Bacillus cereus* (diarrheal type) after consumption of meat.
4. Wound infections: *Bacillus subtilis* (most of times the laboratory contaminants).

Media

1. Blood agar
2. Gelatin stab agar—*B. anthracis* (shows inverted fir tree appearance colony).
3. Selective medium—MYPA (mannitol egg yolk polymyxin phenol red agar) for *B. cereus.*

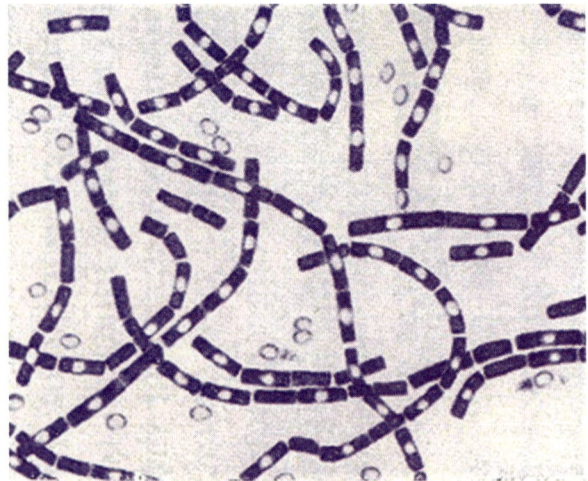

Fig. 3.9: Gram-positive bacilli with non-bulging spore

FAQs

1. Anthrax toxin: Plasmid encoded
2. Bacillus anthrax capsule—**polypeptide** in nature; made up of **polyglutamate**.

Clostridium Species

Gram-positive to Gram variable, highly pleomorphic (3–8 µm × 0.4–1.2 µm), anaerobic, **bulging spore** (Fig. 3.10).

Gram staining: Gram-positive **large** bacilli with **bulging spores**.

Characteristics on Gram staining:

1. *Clostridum perfringens*: Box car shaped bacilli with straight parallel sides and subterminal-spores seen on culture, only this *Clostridium* species is capsulated and non-motile.
2. *Clostridium tetani*: Spherical, terminal spore called as drumstick appearance.
3. *Clostridium tertium*: Oval, terminal spore called as tennis-racket appearance.
4. *Clostridium bifermentans*: Oval, **central** spores.

Diseases caused

1. Gas gangrene: *Clostridium perfringens, Cl. novyi, Cl. septicum*

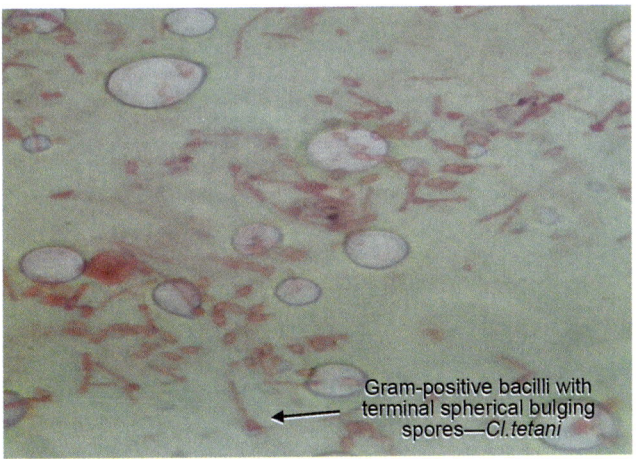

Gram-positive bacilli with terminal spherical bulging spores—*Cl.tetani*

Fig. 3.10: Gram-positive bacilli with bulging spore

2. Food poisoning: *Clostridium perfringens, Clostridium botulinum*
3. Tetanus: *Clostridium tetani*
4. Pseudomembranous colitis: *Clostridium difficile*

Anaerobic media

1. Robertson's cooked meat broth (RCM) and thioglycollate broth
2. Pre-reduced anaerobically sterilized medium (PRAS)
3. *Selective medium*: Neomycin blood agar, kanamycin-vancomycin blood agar, bacteroides bile esculin (BBE) agar, cycloserine cefoxitin fructose agar (CCFA)

Other methods of maintaining anaerobiasis

1. McIntosh and Filde's anaerobic jar.
2. Gas pak system (Fig. 3.11) and Anaeropacks.
3. Anaerobic glove box.

FAQs

1. Examples of obligate anaerobes: *Cl. hemolyticum, Cl. novyi, Cl. perfringens.*
2. Examples of Aerotolerant anaerobes *Cl. histolyticum, Cl. tertium, Cl. carnis*
3. Examples of non-sporing Gram-positive bacilli (non-sporing anaerobes): *Propionibacterium* sp, *Bifidobacterium* sp, *Eubacteria, Arcanobacterium* and *Bacteroides.*

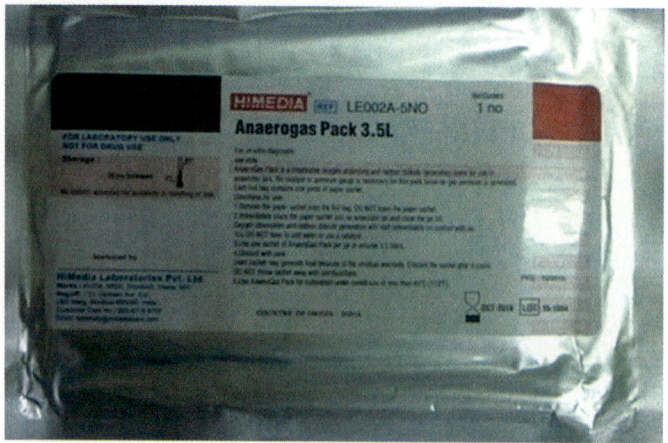

Fig. 3.11: Gas pack

4. Examples of anaerobic Gram-positive cocci: *Peptococcus, Peptostreptococcus, Sarcina, Anaerococcus.*
5. Examples of anaerobic Gram-negative cocci: *Veillonella, Anaeroglobus, Anaerosphaera.*
6. Toxins produced by *Cl. tetani*: Tetanospasmin and tetanolysin.
7. Target hemolysis: *Cl. perfringens.*
8. *Clostridium* species showing swarming: *Cl. tetani* except type 6.

ACID-FAST BACILLI (Fig. 3.12)

Examples

1. *Mycobacterium tuberculosis*
2. *Non-tuberculous Mycobacterium*
3. *Mycobacterium leprae*

Mycobacterium Tuberculosis (MTB)

Diseases caused

1. Pulmonary tuberculosis
2. Extrapulmonary tuberculosis—tubercular meningitis, lymphadenitis, pleural effusion, genitourinary tuberculosis, abdominal kochs and miliary or disseminated tuberculosis.

Media

1. *Solid media*:
    1. Egg-based medium—Lowenstein-Jensen medium (LJ medium) and Petragnani medium.
    2. Agar-based medium—Middlebrook 7H10 and 7H11.
    3. Blood-based medium—Tarshis medium.
    4. Potato-based medium—Powlowsky's medium.
    Colony characteristics on LJ medium: For *M.Tb* colonies are—rough, tough, buff-colored and non-pigmented.

2. *Liquid media*:
    1. Middlebrook 7H9 and 7H12 medium
    2. Dubo's oleic acid medium
    3. Sula's and Sauton's medium
    4. Proskauer and Beck's medium.

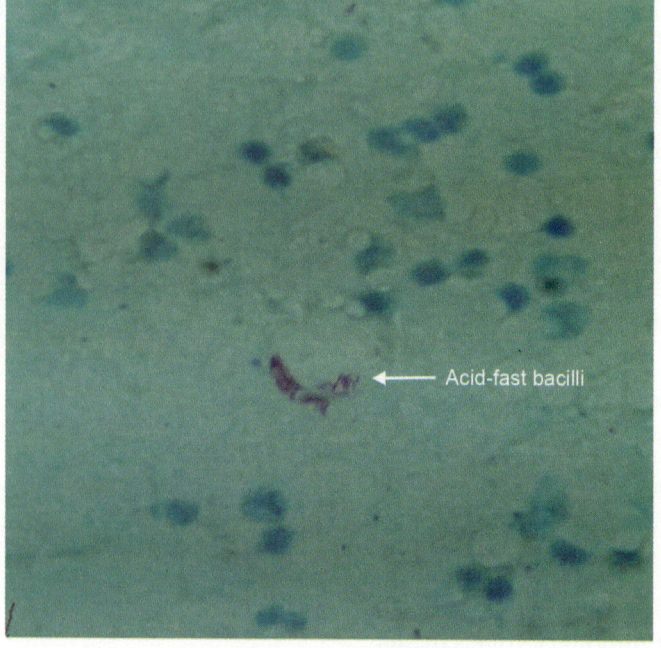

Acid-fast bacilli

Fig. 3.12: Acid-fast bacilli

FAQs

MTB complex includes:

1. *M. tuberculosis*
2. *M. bovis*
3. *M. caprae*
4. *M. africanum*
5. *M. microti*
6. *M. pinnipedii*

Samples

Sputum sample:

i. As per Revised National Tuberculosis Control Programme (RNTCP), to diagnose pulmonary tuberculosis, two sputum samples are recommended of which one spot and other early morning sample.

ii. Sputum samples are subjected to concentration decontamination procedure using NALC-NaOH (N-acetyl l-cysteine +2% NaOH), the most commonly used method. Other method used is the Petroff's method (4% NaOH).

iii. Ziehl-Neelsen (ZN) staining—**acid-fast bacilli (3–4 μm × 0.2–0.8 μm), beaded, serpentine cord-like arrangement**
- **Decolorizer used**—25% H_2SO_4 as per RNTCP.
- **ZN smear comes positive, if number of bacilli in sputum sample is >10,000/ml.**

RNTCP grading of sputum samples

No. of bacilli	Grading	Number of fields scanned
No AFB in 100 fields	Nil	100
1–9/100 hpf	Scanty	100
10–99/hpf	1+	100
1–10/hpf	2+	50
>10/hpf	3+	20

Hpf—high power field (oil immersion lens)

Non-tuberculous Mycobacterium (NTM)/ Mycobacterium other than Tuberculosis (MOTT)

Classification of NTM is called as Runyon's classification.

Diseases caused

1. *Pulmonary disease*: Pleuritis, bronchitis, interstitial pneumonia caused by *M. avium complex, M. kansasii.*
2. *Skin and soft tissue infection*:
   - *M. marinum* causes swimming pool granuloma
   - *M. ulcerans* causes Buruli ulcer
3. *Cervical lymphadenitis*: *M. avium* complex, *M. scrofulaceum* causing scrofula.
4. *Disseminated disease*: *M. avium* complex, *M. kansasii.*

Colony characteristics on LJ medium: For non-tuberculous mycobacteria—colonies are smooth, moist and easily emulsifiable, and pigmented in the case of photochromogens.

FAQs

1. Sample inoculation is done in biosafety cabinet (BSC type II for sample processing and BSC type III for culture processing).

2. Modified ZN staining also called as cold staining or Kinyoun's method is also used for staining the tubercular bacilli. It is different from original ZN staining as here instead of heating the concentration of phenol used while making Carbol fuchsin is high (9%). Other advantage of the procedure is less toxic fumes generation.

3. Modified ZN staining is also used for staining other acid-fast organisms but concentration of H_2SO_4 used is different in each:

   1. *Cryptosporidium sp, Cyclospora cayetenensis*—10% H_2SO_4
   2. *Spores*—0.2–0.5% H_2SO_4
   3. *Nocardia*—1% H_2SO_4

4. Most commonly used fluorescent dye for staining the tubercle bacilli is auramine where bacilli appears as bright brilliant green against a dark background.

5. Different biochemical reactions positive for *M. tuberculosis* are:

   1. Thermostable catalase negative
   2. Niacin positive
   3. Nitrate reduction test positive
   4. Pyrazinamidase test positive
   5. TCH resistant (TCH—thiophene carboxylic acid hydrazide)

6. Bottles used for LJ medium are called as Mc Cartney Bottle.

7. Antibiotic susceptibility testing for *M. tuberculosis* is done by 1% proportion method using LJ-medium with antibiotics like streptomycin (S), rifampicin(R), isoniazid (H), ethambutol (E) except pyrazinamide.

8. Multidrug resistant tuberculosis (MDR-TB): Defined as tubercle bacilli resistant to isoniazid and rifampicin with or without resistance to other first-line antitubercular drugs.

9. Extremely drug resistant tuberculosis (XDR-TB): Tuberculous bacilli which is resistant to isoniazid, rifampicin, fluoroquinolones (ofloxacin/levofloxacin) and at least one injectable aminoglycoside (kanamycin, amikacin and capreomycin).

10. DOTS (directly observed treatment short course): A community based treatment and care of TB patient under supervision.

11. STOP TB strategy: Aim is to reduce the global incidence of TB <1/million population/yr.

12. BCG vaccine: Strain used is Danish 1331 prepared in Central BCG laboratory, Guindy, Chennai.

13. Gene-Xpert: Fully automated machine based on PCR principle to detect pulmonary and extrapulmonary tuberculosis and also resistance of tuberculous bacilli to rifampicin.

14. MGIT (mycobacterium growth indicator tube): Non-radiometric method for the detection of tuberculous bacilli.

15. Drug used in chemoprophylaxis for all asymptomatic contacts of a smear positive case is isoniazid.

Mycobacterium Leprae

Diseases caused

Leprosy: Tuberculoid leprosy, borderline leprosy and lepromatous leprosy.

Media

1. Not cultivable in any artificial culture media.
2. Leprosy bacilli can be grown in foot pads of mice and nine banded armadillo called as *Dasypus novemictus*.

FAQs

1. *Samples*:
   - Split skin smears—7 samples are collected to diagnose leprosy—four from skin (buttock, forehead, chin, cheek), one from nasal septal mucosa and one each from both ears.
   - Other samples are biopsy from nodular lesions, thickened nerves and lymph nodes.
2. *Ziehl-Neelsen (ZN) staining*: Straight to slightly curved acid-fast bacilli seen (1–8 μm × 0.2–0.5 μm) arranged singly or as

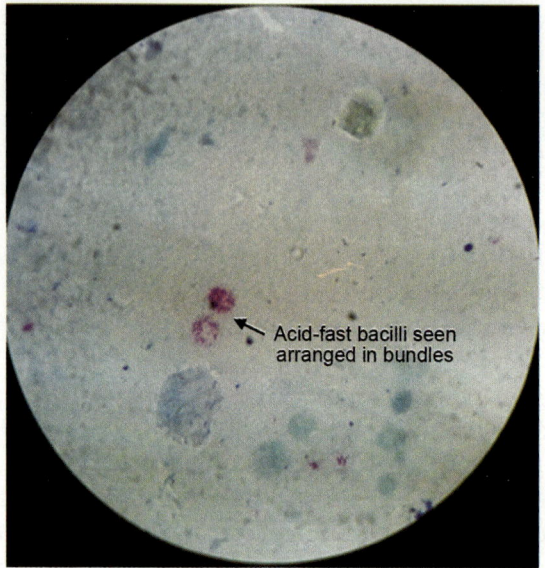

Fig. 3.13: Acid-fast bacilli in bundle

groups, intracellularly in form of *bundles called as globi*. This is also called as Virchow's cells or foamy cells.

- **Decolorizer used:** 5% H_2SO_4.
- **Grading of smears:**

No. of bacilli	Grading
1–10/100 hpf	1+
1–10/10 hpf	2+
1–10/hpf	3+
10–100/hpf	4+
100–1000/hpf	5+
>1000/hpf or bacilli in clumps	6+

Hpf—high power field (oil immersion lens)

3. *Bacteriological index*: Total grading of individual smears/ number of smears examined.
4. *Morphological index*: Percentage of uniformly stained bacilli/ total number of bacilli counted.
   - It is considered as an indicator to monitor treatment response.

- Best indicator to monitor treatment response—SFG % which is called as the solid fragmented granular rod percentage.

5. *Lepromin test*: Delayed type of hypersensitivity response appears when 0.1 ml of lepromin-A antigen injected intradermally to forearm. It shows:

   1. Fernandez reaction (early reaction): Induration of ≥10 mm seen at 24–48 hrs.
   2. Mitsuda reaction (late reaction): Nodule of ≥5 mm seen after 21 days is considered as positive test. **This reaction is more important than early reaction.**

   Use: Classify leprosy lesions, determine prognosis of patient on treatment and to access the resistance to leprosy in individuals.

6. *Serology*: Antigen of lepra bacilli used to detect antibodies in the patient is phenolic glycolipid-1(PGL-1).

7. *Drugs used in treatment*:
   i. Paucibacillary leprosy: Dapsone and rifampicin
   ii. Multibacillary leprosy: Dapsone, rifampicin and clofazimine.

BUDDING YEAST CELLS (Fig. 3.14)

Gram-positive, small (2–4 µm), thin-walled, ovoid budding yeast cells (blastoconodia) seen.

Species of *Candida*: *C. albicans*, *C. dublinensis*, *C. glabrata*, *C. guilliermondii*, *C. kefyr*, *C. krusei*, *C. lusitaniae*, *C. tropicalis*.

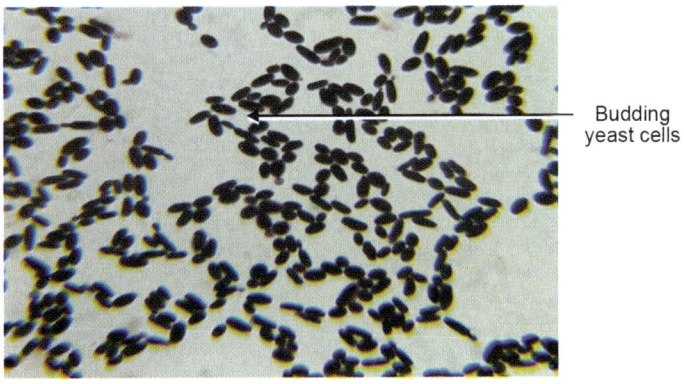

Budding yeast cells

Fig. 3.14: Budding yeast cell

Diseases caused

1. *Mucocutaneous candidiasis*: It includes oral thrush, esophagitis, vaginitis, cutaneous candidiasis (folliculitis, balanitis, intertrigo, paronychia and onychomycosis).
2. Deep-seated candidiasis includes central nervous system, respiratory tract, endocarditis, urinary tract candidiasis, candida arthritis, osteomyelitis, costochondritis and myositis.

Media used

1. Sabouraud dextrose agar plain.
2. Sabouraud dextrose agar with antibiotics—cyclohexamide and chloramphenicol.
3. Corn meal agar—used for species identification of *Candida* species.

   Colony—small, white to cream, smooth or slightly wrinkled.

FAQs

1. *Germ tube test* (Figs 3.15 and 3.16): A method to make a presumptive identification of *Candida albicans*.
   i. Positive—*Candida albicans*.
   ii. Negative—*Candida tropicalis, C. krusei, C. kefyr.*

Note: The presence of germ tube is indicative of *Candida albicans* or *Candida dublinensis*, which are further differentiated with the help of corn meal agar.

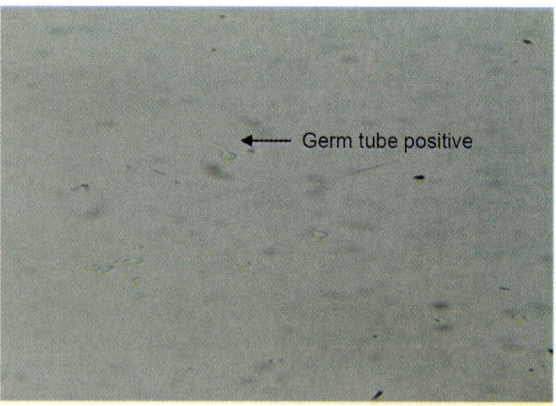

← Germ tube positive

Fig. 3.15: Germ tube test positive

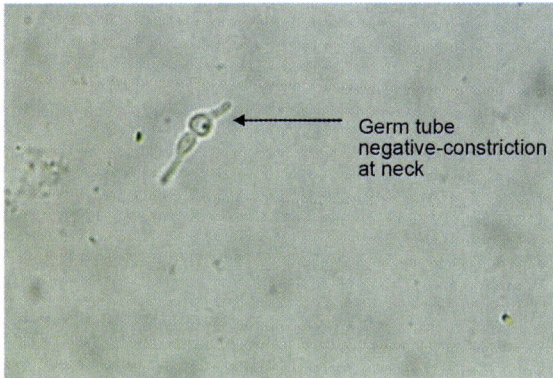

Germ tube
negative-constriction
at neck

Fig. 3.16: Germ tube test negative

ASPERGILLUS (Fig. 3.17)

Lactophenol cotton blue (LPCB) mount showing hyaline septate hyphae ending at globular vesicle. On vesicle are present phialides. The first row of phialides is also called as metulae. These phialides bear the chains of conidia.

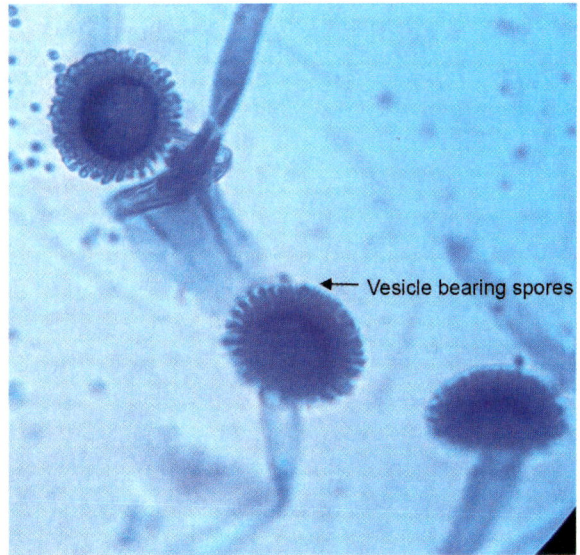

← Vesicle bearing spores

Fig. 3.17: *Aspergillus* spp LPCB mount

Species of Aspergillus: A. fumigatus, A. flavus, A. niger, A. terreus, A. nidulans.

Diseases caused

1. Allergic bronchopulmonary aspergillosis
2. Fungal ball
3. Sinusitis
4. Otomycosis
5. Keratitis
6. Invasive pulmonary aspergillosis

Media used

Sabouraud dextrose agar plain:

- Sabouraud dextrose agar with antibiotics—cyclohexamide and chloramphenicol. This medium does not support the growth of *Aspergillus* spp.
- Colony—smoky-green, velvety to powdery in nature.

FAQs

1. Drug of choice for the invasive aspergillosis: Voriconazole.
2. Serological: ELISA—based on detection of Galactomannan antigen in patient serum or urine.

India ink preparation (Fig. 3.18)

- India ink and Nigrosin dye are used to demonstrate some capsulated organisms like *Streptococcus pneumoniae, Klebsiella pnemoniae* and *Cryptococcus neoformans.*
- *Principle of the test*: India ink or Nigrosin dye cannot penetrate capsular material thus background of the organism is stained black and organism appears as unstained structure.
- *Method of test*: Small amount of growth or direct clinical sample (in case of meningitis CSF) is mixed with one drop of India ink or Nigrosin on a glass slide then covered with a cover slip and observed under low or high power objective lens.

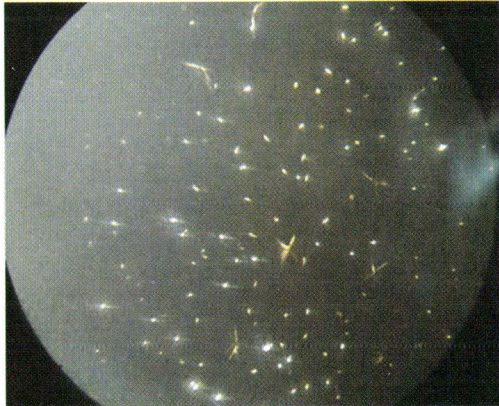

Fig. 3.18: India ink preparation – *Cryptococcus* spp

RHINOSPORIDIUM SPHERULES (Fig. 3.19)

Slide shows spherules, i.e. sporangium (10–200 µm) containing endospores of size 6–7 µm.

Causative agent: *Rhinosporidium seeberi* (protistan parasite).

Diseases caused
Large friable polyps: Painless localized infection of mucous membranes typically involving nose, upper airways and conjunctiva.

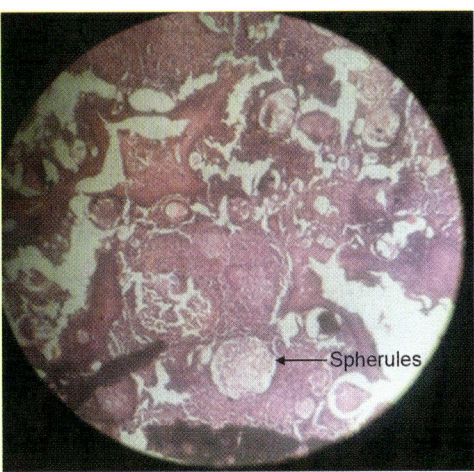

Spherules

Fig. 3.19: Rhinosporidium

Media

Cannot be cultured.

FAQs

1. Rhinosporidiosis is most common in southern India (Tamil Nadu, Kerala, Orissa and Andhra Pradesh) and Sri Lanka.
2. *Mode of infection*: Contact with contaminated water while swimming
3. *Treatment of choice*: Surgery
   - Other organisms which cannot be cultivated on artificial culture media or cell lines are: *M. leprae, Lacazia loboi, Molluscum cotagiosum* virus.

MOLLUSCUM CONTAGIOSUM (Fig. 3.20)

H and E stained tissue section showing nestles of 20–30 µm intracytoplasmic acidophilic inclusion bodies displacing the nucleus, also called as molluscum bodies or HP bodies (Henderson–Peterson bodies).

It belongs to:

- *Family*: Poxviridae
- *Genus*: Mollscipoxvirus—a double-stranded DNA (dsDNA) virus.

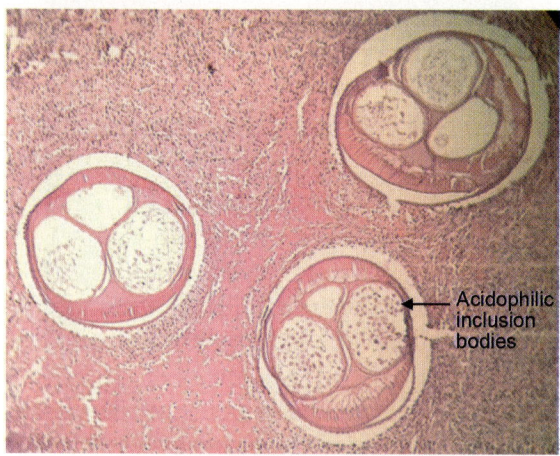

Fig. 3.20: Molluscum contagiosum

Diseases caused

1. A benign self-limited skin tumor.
2. *Papular eruptions*: Small, smooth, dome-shaped, pearly fleshy colored nodule, often umblicated.
   Cell lines used: Cannot be cultured.

FAQs

1. *Mode of transmission*:
   i. Sexual route
   ii. Indirect contact—Fomites (towel sharing), barbers, etc.
2. Method for confirmation of diagnosis is electron microscopy showing the characteristic brick-shaped virions.
3. *Treatment*: The contents of lesion are exposed and treated with 25% phenol solution or 30% trichloroacetic acid.

OOCYST OF CRYPTOSPORIDIUM (Fig. 3.21)

Modified ZN staining (10% H_2SO_4) showing spherical to ovoid, 4–5 µm in diameter acid fast oocyst.

Diseases caused

Diarrheal infection in AIDS patients.

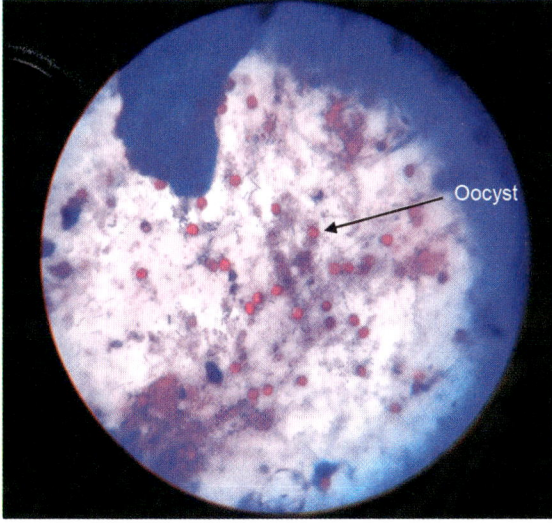

Fig. 3.21: Cryptosporidium oocyst

FAQs

1. Mode of transmission—feco-oral route.
2. Oocyst contains—4 sporozoites without sporocysts.
3. Other sporozoa causing oppurtunistic diarrheal infection in AIDS patients are—*Cystoisospora belli* (previously *Isospora belli*), *Cyclospora cayetanenesis, Mirosporidium*.

FEMALE GAMETOCYTE OF PLASMODIUM FALCIPARUM

1. Thin peripheral blood smear showing the presence of female gametocytes of *P. falciparum*, size approx 1–12 µm × 2–3 µm, crescent-shaped with pointed ends, central compact nucleus and surrounded by malarial pigment. Infected RBC membrane is stretched on the gametocyte (Fig. 3.22).
2. *Other features of* **P. falciparum** *infection:*
   1. Infect RBCs of all ages
   2. Ring forms in RBCs—multiple rings in individual RBC and "Accole" forms
   3. Maurer's clefts in RBCs
   4. Banana-shaped gametocytes

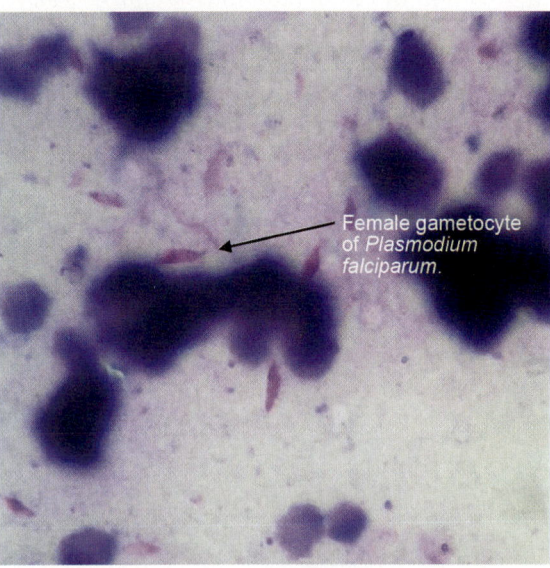

Fig. 3.22: Sickle-shaped gametocyte of *P. falciparum*

Diseases caused
1. Malaria
2. Complication of *P. falciparum* infection is pernicious anemia which is a complex of cerebral malaria, algid malaria and septicemic malaria.
3. Blackwater fever—manifestation of **repeated infection** with *P. falciparum* in patients who were inadequately treated with quinine.

Media used

Trager and Jensen culture medium: Parasite is maintained in continuous culture in human RBC with RPMI – 1640 (Rosewall Park Memorial Institute) + 7% CO_2 + 1–5% O_2. This culture is used for antigen production only.

FAQs
1. Intermediate host—man
2. Definitive host—female anopheles mosquito
3. Inside the liver (pre-erythrocytic/primary exoerythrocytic schizogony)—no clinical symptom and no pathological damage.
4. Erythrocytic schizogony—cause of malarial paroxysm
   Note: *P. vivax* and *P. ovale*—exo-erythrocytic (secondary exo-erythrocytic schizogony) shows **latent hypnozoites** which are the cause of relapse.
5. Species of *Plasmodium* infecting humans:
   1. *Plasmodium falciparum*: Malignant tertian fever
   2. *P. vivax*: Benign tertian fever
   3. *P. ovale*: Ovale tertian
   4. *P. malariae*: Quartan fever
   5. *P. knowlesi*: Quotidian fever
6. *Thick smear*: During preparation—RBCs lysed by distilled water whereas the intact parasites remain and are concentrated.
   Use:
   i. Detecting parasites
   ii. Detecting malarial pigments

7. *Thin smear*: 200–300 fields are examined before smears are considered as negative.

   Use:
   i. Examined for species diagnosis
   ii. Estimation of parasitemia
   iii. RBC morphology
   iv. Diagnose mixed infections.

8. Different stains used for staining malaria parasite in the peripheral blood smear:
   1. Giemsa stain—most commonly used method for both thin and thick smears
   2. Field stain
   3. Leishman stain
   4. Jaswant Singh Battacharya—standard method used under National Eradication Program in India.

9. *Other methods to diagnose malaria*: Quantitative buffy coat (QBC) based on principle of using acridine orange (AO) which stains nucleic acid in parasite at buffy coat layer, i.e. RBC–WBC interface.

10. *Treatment*: Chloroquine, mefloquine, artemisinin combination therapy (ACT).

SCOLEX OF TAENIA SAGINATA/BEEF TAPEWORM/UNARMED TAPEWORM OF MAN (Fig. 3.23)

Scolex: Large, quadrate, with 4 suckers (may be pigmented), rostellum and hooklets are absent.

FAQs
1. *Definitive host*: Man
2. *Intermediate host*: Cattle
3. Larval stage called as cysticercus bovis is present in cows only.

SCOLEX OF TAENIA SOLIUM/PORK TAPEWORM/ARMED TAPEWORM OF MAN

Scolex: Small, globular with 4 suckers (not pigmented), rostellum present is armed with a double row of 25–30 alternating large and small hooklets.

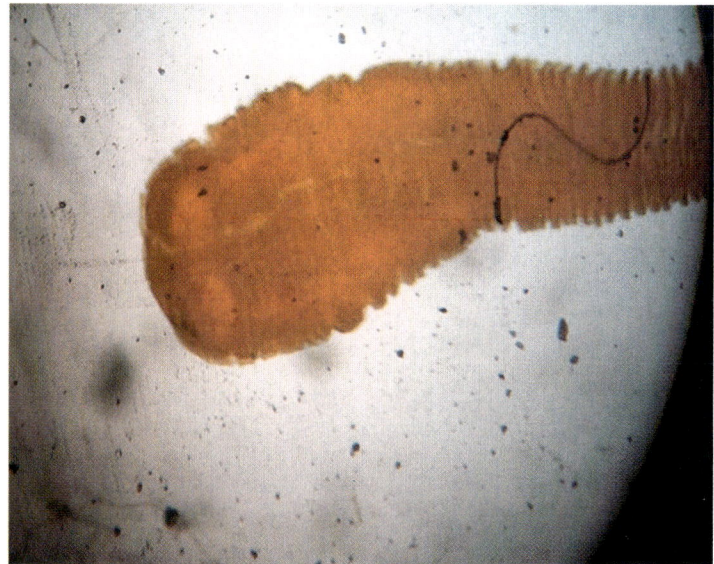

Fig. 3.23: Scolex of *Taenia* spp

Rostellar hooklets: Shaped like daggers.

FAQs
1. *Definitive host*: Man
2. *Intermediate host*: Pig
3. Larval stage called as cysticercus cellulosae is present in pig and may develop in man.

GRAVID SEGMENT OF TAENIA SOLIUM/PORK TAPEWORM/ ARMED TAPEWORM OF MAN (Fig. 3.24)

- *Length*: 12 mm
- *Breadth*: 6 mm
- *Number of lateral branches*: 5–10
- Gravid segments are passed in feces in chains of 5–6.

FAQs
1. *Definitive host*: Man
2. *Intermediate host*: Pig.

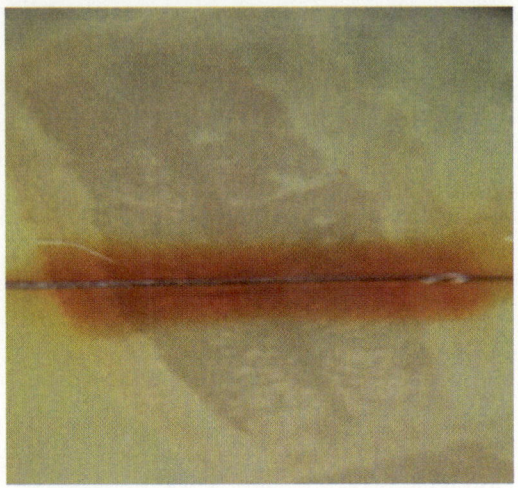

Fig. 3.24: Gravid segment of *Taenia* spp

ECHINOCOCCUS (ADULT WORM) (Fig. 3.25)

Other name: Dog tapeworm/hydatid worm.

Adult worm: 3–6 mm in length, it can be divided into:

1. Scolex—pyriform in shape, 300 μm in diameter, 4 suckers and a protruding rostellum with 2 circular rows of hooklets.

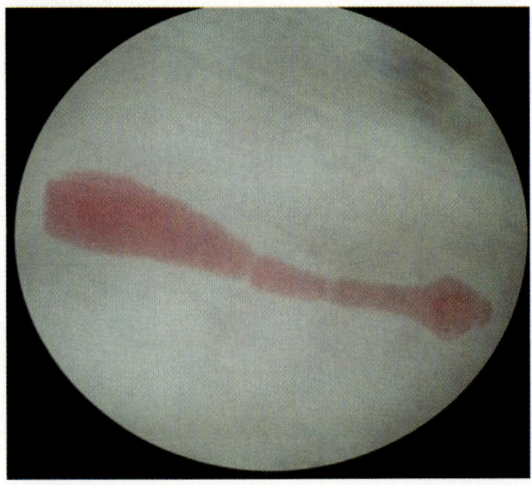

Fig. 3.25: Segments of *Echinococcus granulosus*

2. Neck
3. Strobila—consists of **3 segments** (immature, mature and gravid segment).

FAQs

1. *Definitive host*: Dog
2. *Intermediate host*: Sheep and man
3. *Lifespan of adult worm*: Short (about 6 months)
4. Species of *Echinococcus* infecting humans:
   1. *E. granulosus*: It causes hyadatid disease/hydatid cyst.
   2. *E. multilocularis*: It causes alveolar echinococcosis.
   3. *E. vogeli*: It causes polycystic hydatid disease.
5. *Casoni test*: Type I hypersensitivity reaction seen when 0.2 ml of antigen is injected intradermally lead to wheal formation ≥5 cm in 30 minutes.
6. *Hydatid fluid*: Clear, colorless or pale yellow, slighty acidic fluid secreted by the endocyst or inner germinal layer of the hydatid cyst.
7. *Hydatid sand*: Centrifuged deposit of hydatid fluid shows hydatid sand which consists of brood capsules, free scolices and hooklets.

HYDATID CYST

1. **Pericyst**: Shows the presence of fibroblasts and new blood vessels and parasite derives nourishment through it.
2. **Ectocyst** (Fig. 3.26): Up to 1 mm thick outer laminated hyaline membrane, elastic and when incised/ruptured, it coils on itself exposing inner germinal layer.
3. **Endocyst (inner germinal layer)**: Very thin, 22–25 μm cellular layer, consists of brood capsules and daughter cysts and secretes specific hydatid fluid (Fig. 3.27).

FAQs

1. *Development of hydatid cyst*: Very slow, 4 cm by end of one year.
2. *Distribution site of hydatid cyst*:
   • Ist most common site—liver
   • IInd most common site—lung.

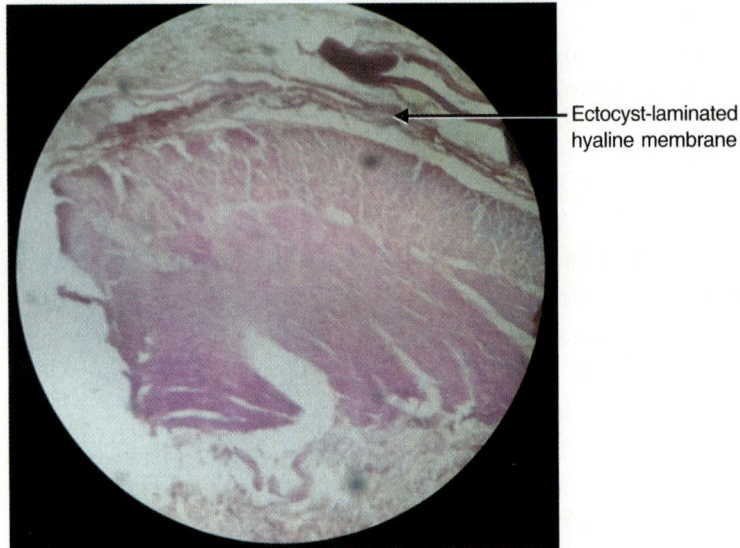

Ectocyst-laminated hyaline membrane

Fig. 3.26: Ectocyst of *Echinococcus*

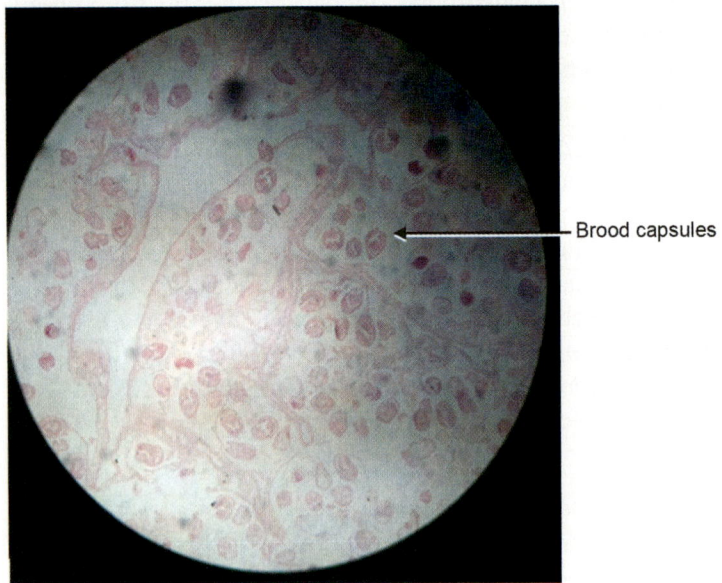

Brood capsules

Fig. 3.27: Brood capsule of *Echinococcus*

ANCYLOSTOMA DUODENALE (ADULT WORM)

Slide shows the characteristic buccal capsule and esophagus of adult worm.

The anterior end is curved dorsally in the same direction as body curvature.

Other name: Old world hookworm.

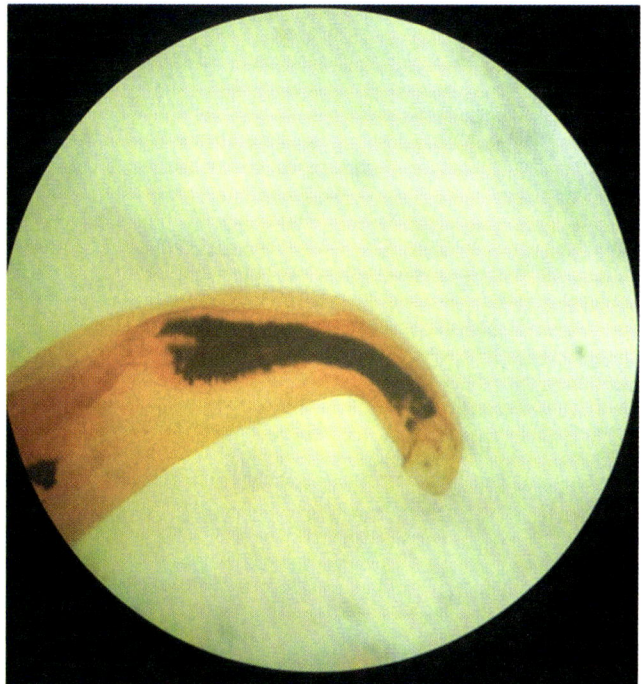

Fig. 3.28: Head of hookworm

FAQs

1. *Single host*: Man.
2. *Adult female worm* is 9–13 mm × 0.6 mm, tapered posterior end and genital pore opens at the junction of the middle and posterior third of the body.
3. *Adult male worm* is 5–11 mm × 0.4–0.5 mm, posterior end is expanded in an umbrella-like fashion called as copulatory bursa, genital pore opens posteriorly with cloaca.

4. *Hookworms infecting humans*:

 i. *Ancylostoma duodenale* causes *Ancylostoma* dermatitis or ground itch, pulmonary lesions like bronchitis and bronchopneumonia and iron deficiency anemia.

 ii. *Necator americanus causes* iron deficiency anemia.

5. **Difference between *Ancyclostoma duodenale* and *Necator americanus*:**

Ancyclostoma duodenale	*Necator americanus*
Larger and thicker	Smaller and thinner
Anterior end bends in the same direction as body curvature	**Anterior end** bends in the opposite direction as body curvature
6 Teeth—4 hook-like on ventral surface and 2 knob-like on dorsal surface	**4 cutting plates**—2 each on ventral and dorsal sides
Copulatory bursa has 13 rays and 2 separate spicules	**Copulatory bursa** has 14 rays and 2 spicules fused at tip
Posterior spine present	**Posterior spine** absent
More pathogenic	Less pathogenic

6. *Cutaneous larva migrans*: Causes creeping eruptions leading to formation of itchy red lines or tracks under the skin surface. It is caused by:

- *Ankylostoma braziliensis*
- *Ankylostoma caninum*
- *Ganathostoma* sp
- *Uncinaria stenocephala*
- *Bunostomum phlebotomum*.

ENTEROBIUS VERMICULARIS (EGG)

Other name: Threadworm/pinworm/seatworm.

Slide shows: Colourless, non-bile-stained, **planoconvex** (flattened on one side) egg, measuring 60 × 30 µm and surrounded by a thin smooth transparent shell usually containing a developed larvae (Fig. 3.29).

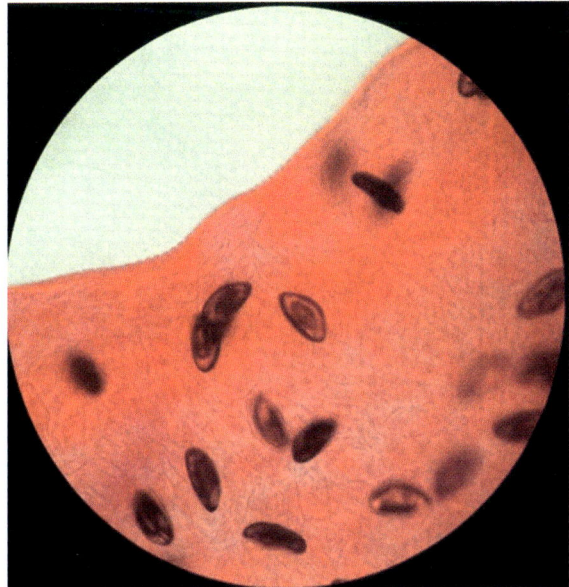

Fig. 3.29: Egg of *Enterobius vermicularis*

FAQs

1. *Single host*: Man
2. *Habitat*: Cecum and appendix
3. *Enterobius* species infecting humans:
    i. *E. vermicularis*
    ii. *E. gregorii*
4. *NIH swab*: National institute of health swab consists of glass rod with a cellophane at one end to collect the eggs deposited in perianal and perineal skin.
5. Scotch cellophane tape method also used to collect the sample.

MICROFILARIA (Fig. 3.30)

Slide shows: Sheathed microfilaria measuring 245–295 µm 7.5–10 µm. Hyaline sheath covering the microfilaria is longer (359 µm). The somatic cells and nuclei appear as granules in the central axis of microfilaria. The tail tip is free from nuclei.

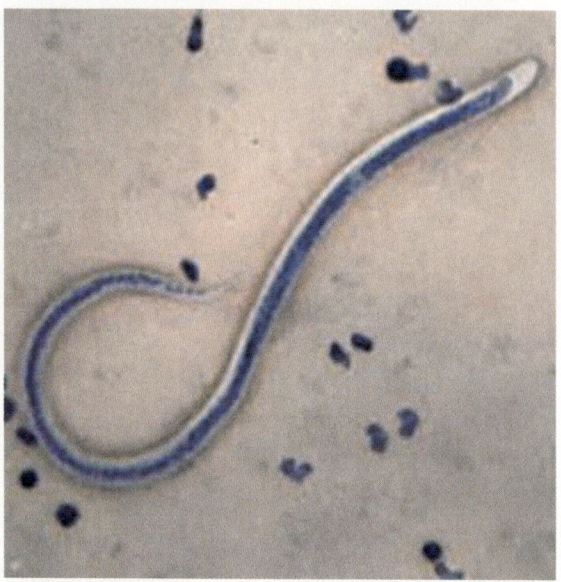

Fig. 3.30: Microfilaria of *Wuchereria bancrofti*

FAQs

1. *Definitive host*: Man.

2. *Intermediate host*: Female anopheles belonging to genera *Culex, Aedes* and *Anopheles*.

3. *Habitat*: Lymph nodes and lymphatics (usually inguinal, scrotal and abdominal) of man.

4. *Filarial nematodes infecting humans*:

   i. *Wuchereria bancrofti*—causes lymphangitis, lymphatic obstruction, lymphedema and elephantiasis.

   ii. *Brugia malayi*—causes filariasis

   iii. *Loa loa* (eyeworm)—causes fugitive or calabar swellings which are a painless edematous swellings, conjunctival granulomata.

   iv. *Onchocerca volvulus* (convoluted filarial)—causes river blindness.

5. Sheathed microfilaria seen in: *Wuchereria bancrofti, Brugia malayi* and *Loa-loa*.

6. Unsheathed microfilaria seen in: *Mansonella perstans, Mansonella ozzardi, Mansonella streptocerca,* and *Onchocerca volvulus.*

7. Microfilaria present in blood: *Wuchereria bancrofti, Brugia malayi* and *Loa-loa* and *Mansonella perstans, Mansonella ozzardi* species.

8. Microfilaria present in skin: *Mansonella streptocerca* and *Onchocerca volvulus.*

Organisms—vector associated

1. *Wuchereria bancrofti*: *Culex quinquefasciatus* (mainly) and *Anopheles* sp.

2. *Brugia malayi*: Mansonia, Anopheles and Aedes.

3. *Loa loa* (eyeworm): Day biting female mango fly (*Chrysops dimidiate*).

4. *Onchocerca volvulus* (convoluted filarial): Day biting female blackfly of genus *Simulium.*

Drug of choice

1. *Wuchereria bancrofti*: Diethylcarbamazine (DEC)

2. *Onchocerca volvulus*: Ivermectin.

ENCYSTED LARVAE OF TRICHINELLA SPIRALIS

Other name: Trichina worm

Slide shows: Encysted larvae in the striated muscle measuring 1×36 µm. The larvae in the cyst are coiled and hence called as spiralis.

FAQs

1. *Definitive host*: Pig (mainly), rats, man is an alternative host.

2. *Intermediate host*: **ABSENT**.

3. *Habitat*: Duodenum and jejunum.

4. *Larvae found in striated muscles*: Deltoid, biceps, gastrocnemius and pectoralis muscle.

5. Minimum number of larvae required to produce symptoms is 100 larvae.

6. Fatal dose is 300,00 larvae.

7. *Drug of choice*: Albendazole.

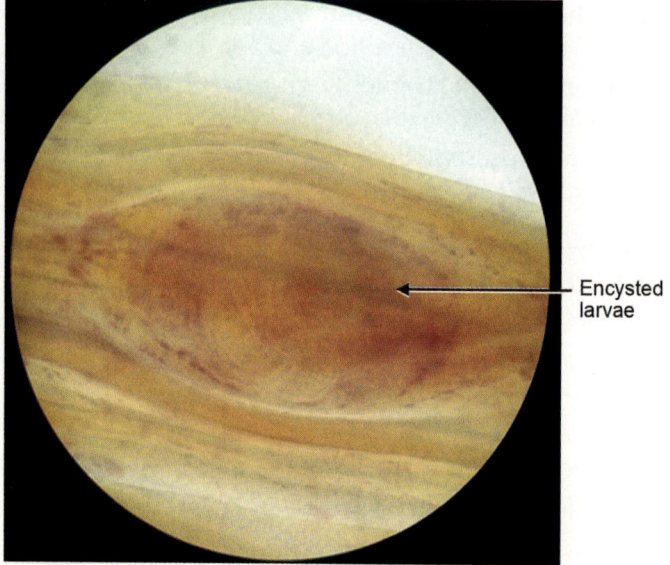

Fig. 3.31: Encysted larva of *Trichinella*

RAT FLEA/XENOPSYLLA CHEOPIS

Slide shows: A small wingless insect, brown in color and characterized by three pairs of legs with body divided into three segments—head, thorax and abdomen.

Fig. 3.32: Rat flea

Diseases transmitted
1. Endemic typhus (murine typhus)
2. Plague
3. Hymenolepsis nana (indirect lifecycle)

BODY LOUSE/PEDICULUS HUMANUS CORPORIS (Fig. 3.33)

Slide shows: A small wingless insect, brown in color, characterized by three pairs of legs and a flattened body divided into three segments—head, thorax (almost square) and abdomen (divided into 9 segments).

Diseases transmitted
1. Epidemic typhus
2. Epidemic form of relapsing fever
3. Trench fever

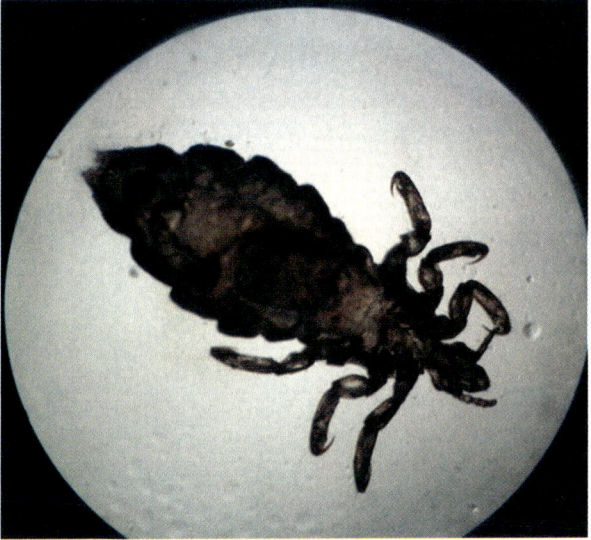

Fig. 3.33: Body louse

ANOPHELES MOSQUITO (Fig. 3.34)

Slide shows: Insect with wings (black and white scales on wings), brown in color, characterized by three pairs of legs and body is divided into three segments—head, thorax and abdomen.

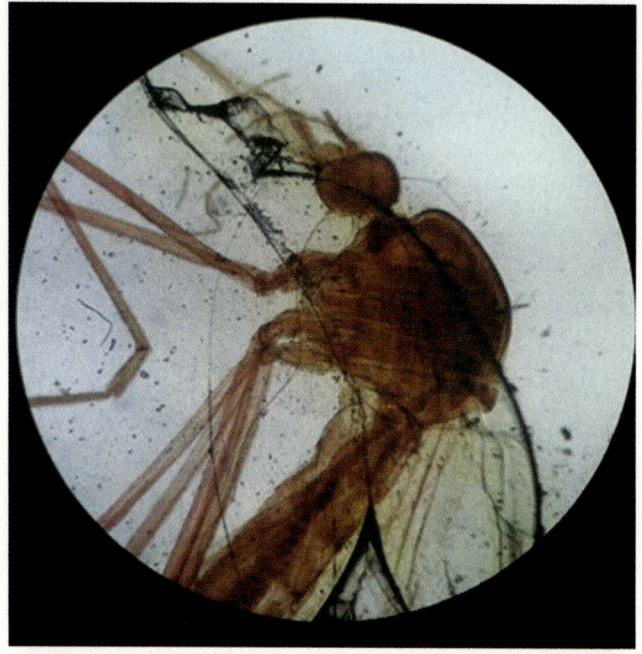

Fig. 3.34: Anopheles mosquito

Head bears: Pairs of long unsegmented antennae, proboscis and two maxillary palps.

Diseases transmitted

1. Malaria
2. Filariasis

MITE (Fig. 3.35)

Slide shows: A small wingless insect (0.55 mm) and body is divided into two main segments—cephalothorax and abdomen.

Diseases transmitted

1. Scabies
2. Scrub typhus
3. Rickettsial pox

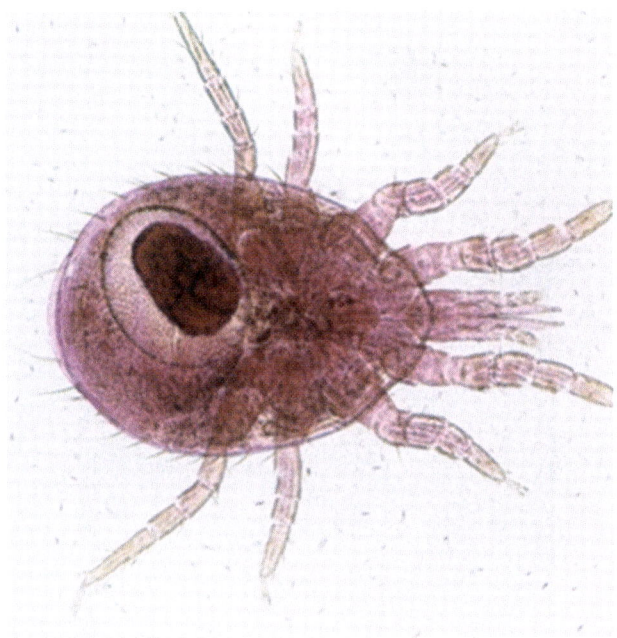

Fig. 3.35: Mite

TICK (Fig. 3.36)

Slide shows: A small wingless insect (5 mm in length), oval, four pairs of legs and body is divided into two main segments—cephalothorax and abdomen.

Diseases transmitted

1. Spotted fever
2. Q-fever
3. Lyme disease
4. Tularemia
5. Arbovirus: Encephalitis, Kyasanur forest disease
6. Babesiosis

Fig. 3.36: Tick

FAQs

1. Hard tick: *Ixodid* ticks
2. Soft ticks: *Argasial* ticks.

REDUVIID BUG

Diseases transmitted: Chagas disease.

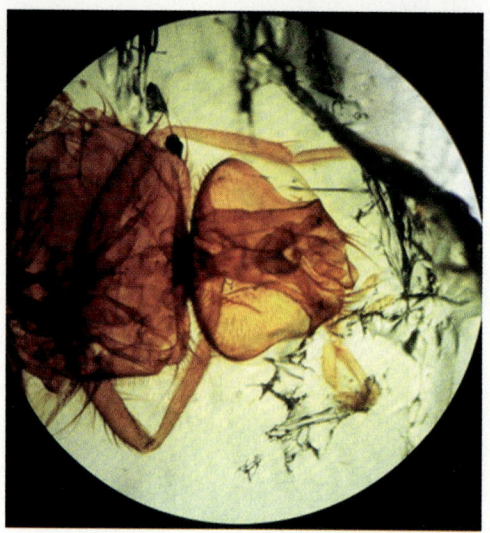

Fig. 3.37: Reduviid bug

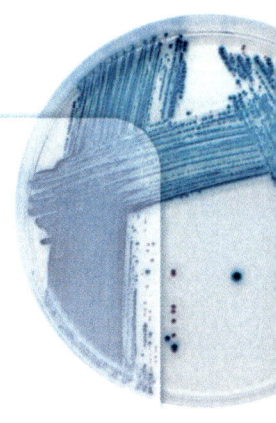

Chapter 4

Spots in Parasitology

Adult Worm of *Taenia* (Fig. 4.1)

Two to three metres in length, segmented worm with body (strobila) composed of chains of proglottids or segment.

Types of proglottids

1. Immature—present near neck
2. Mature—contains male and female genital organs
3. Gravid—primary genital is atrophied but contains uterus filled with eggs.

Fig. 4.1: Adult worms of *Taenia*

Hydatid Cyst

Embryo develops into a fluid-filled bladder called as hydatid cyst (Fig. 4.2).

Fig. 4.2: Hydatid cyst

Hydatid Sand

Centrifuged deposit of hydatid fluid shows hydatid sand (Fig. 4.3) which consists of brood capsules, free scolices and hooklets.

Fig. 4.3: Hydatid sand

Adult Worm of *Trichuris Trichiura*/Whipworm

Specimen is of a large intestine showing the characteristic whip-shaped adult worm of *Trichuris trichiura*. White-colored worm where the anterior 3/5th is very thin and posterior 2/5th is thick and stout resembling handle of whip (Fig. 4.4). The thin anterior end is buried in the mucosa.

Male worms has coiled posterior end.

Female worm has comma or arc-shaped posterior end.

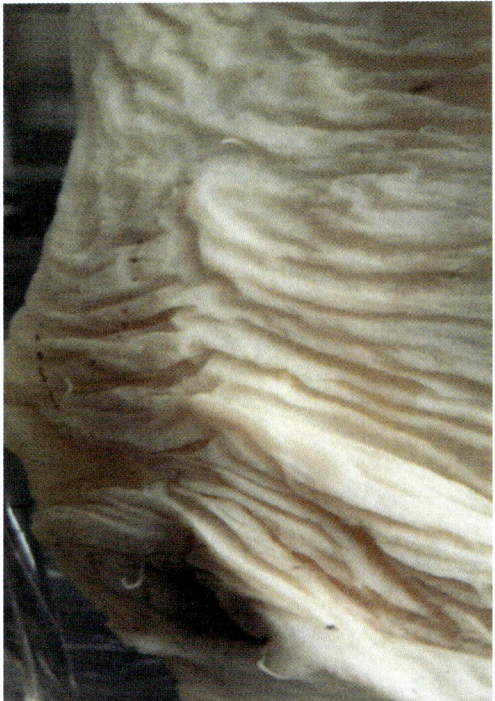

Fig. 4.4: Adult worm of whipworm on colonic mucosa

Adult Worm of *Ascaris*

Specimen showing the characteristic cylindrical adult worm of *Ascaris* (Fig. 4.5). The anterior end is tapered with somewhat less at the posterior end.

Male worm 15–30 cm × 3–4 mm, curved posterior end to form a hook.

Fig. 4.5: Adult worm of *Ascaris*

Female worm 25–40 cm × 5 mm, straight posterior end and shows vulvar waist (vulva opens at junction of anterior one-third and middle one-third on midventral aspects).

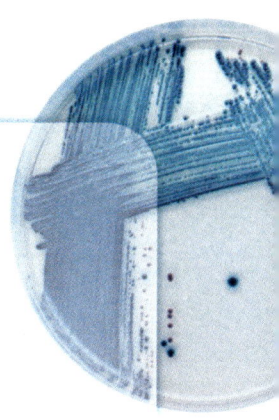

Chapter 5

Instruments

Bunsen Burner (Fig. 5.1)

FAQs
1. Use to sterilize loops, culture tubes, glass slides and scalpels.
2. Blue part of the flame is used for sterilizing.

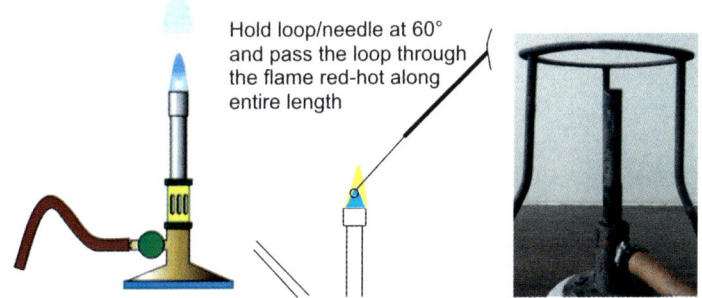

Hold loop/needle at 60°
and pass the loop through
the flame red-hot along
entire length

Fig. 5.1: Bunsen burner

Inoculating Loop (Fig. 5.2)

Types of loop
1. Nichrome wire loop
2. Platinum wire loop

Use
1. To make smears
2. To pick up colonies from culture media and inoculating them.
3. To make bacterial suspensions.

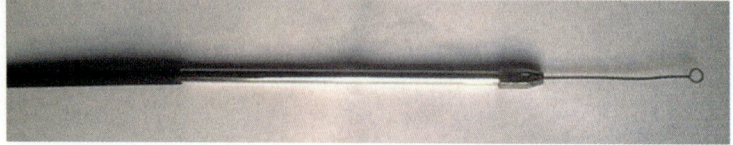

Fig. 5.2: Inoculating loop

Sterilized by: Flaming

Various diameters of loops available are: 1.3 mm, 2 mm and 4 mm.

Straight Wire (Fig. 5.3)

Types of wire
1. Nichrome wire of SWG (standard wire gauge)—26 or 27.
2. Platinum wire (considered as best).

Fig. 5.3: Straight wire

Use
1. To make smears
2. To pick up colonies from culture media and inoculating them by stab culture.
3. To make bacterial suspensions.

Sterilized by: Flaming.

Sterile Universal Containers (Fig. 5.4)

Use: To collect samples—pus, urine, CSF, etc.

Sterilized by—radiation for pre-packed disposable items.

Various sizes available are: 5 ml, 10 ml, 50 ml capacity, etc.

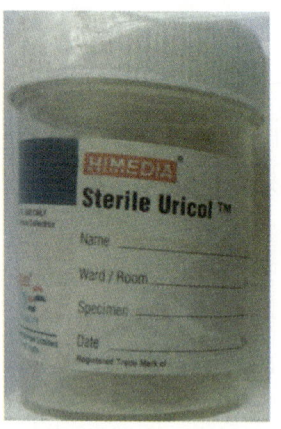

Fig. 5.4: Sterile universal container

Sterile Disposable Syringes (Fig. 5.5)

Use
1. To collect samples—blood samples.
2. To administer injectable drugs.

Fig. 5.5: Sterile disposable syringe

Sterilized by—ionizing radiations like gamma radiations or non-ionizing radiations like infrared rays.

Various sizes available are: 2 ml, 5 ml, 10 ml capacity, etc.

FAQs
1. Best method for the sterilization of **pre-packed disposable syringes**—ethylene oxide gas.
2. Sterilization control for ethylene oxide sterilization—*Bacillus globigii.*

Sterile Swab Sticks (Fig. 5.6)

Use
1. To collect samples from throat, conjunctiva, vagina, etc.
2. To make lawn culture while performing disk diffusion testing.

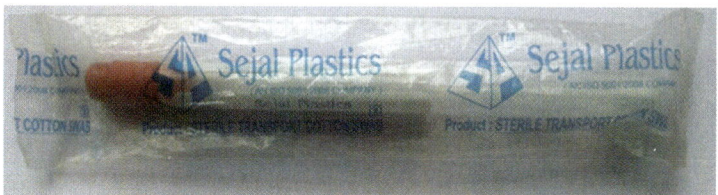

Fig. 5.6: Sterile swab stick

Sterilized by—hot air oven at 160°C for 1 hr after wrapping them in craft paper.

FAQs
1. Cotton swabs are not used to collect suspected gonorrhea infection samples.

Hot Air Oven (Fig. 5.7)

Principle: Based on method of dry heat as sterilization, as it causes protein denaturation, oxidative damage and increases the levels of electrolytes thus finally causing cell death of the organism.

Use: Sterilization of glass wares, swab sticks, oils powder and glycerol.

Sterilized by hot air oven: At 160°C for 1 hr (holding time).

FAQs
Sterility control:
  i. Biological control—strips containing 10^6 spores of the non-toxicogenic strains of *Clostridium tetani* or *Bacillus subtilis*.
 ii. Browne's tube—color of the tube changes from red to green.
iii. Thermocouples.

Fig. 5.7: Hot air oven

Autoclave (Fig. 5.8)

Principle: Based on method of moist heat as sterilization, as it causes protein denaturation and protein coagulation, thus finally causing cell death of the organism.

Use: Sterilization of culture media, aqueous solutions, glass wares, plastic tubes, biohazardous waste.

Sterilized by autoclave

i. At 121°C for 15 min at 15 pound (Ibs) per square inch (psi) pressure.
ii. At 126°C for 10 min at 20 pound (Ibs) per square inch (psi) pressure.
iii. At 133°C for 3 min at 30 pound (Ibs) per square inch (psi) pressure.

FAQs

1. *Sterility control*:
   i. Biological control—strips containing 10^6 spores of the *Geobacillus stearothermophilus*. These spores are killed in 12 min at 121°C.

Fig. 5.8: Autoclave

ii. Browne's tube—color of the tube changes from red to green.

iii. Thermocouples.

2. *Different types of autoclave available*:

i. Horizontal autoclave

ii. Vertical autoclave

iii. Portable (table top) autoclaves—used in outpatient, dental and rural clinics.

3. *Different types of autoclave based on their working*:

i. Gravity displacement autoclave.

ii. High speed prevacuum sterilizers—fitted with a vacuum to ensure the air removal from the sterilizing chamber before the steam is admitted. It works at 134°C with 3–4 min exposure time.

iii. Steam flush—pressure pulsing process—which removes air rapidly by repeatedly alternating a steam flush and a pressure pulse above the atmospheric pressure. It works at 132–135°C with 3–4 min exposure time.

Seitz Filter (Fig. 5.9)

Principle: It is a mechanical method used to remove the microorganism from the substances which are heat-labile like sera, sugars, antibiotic solutions which normally cannot be sterilized by autoclave.

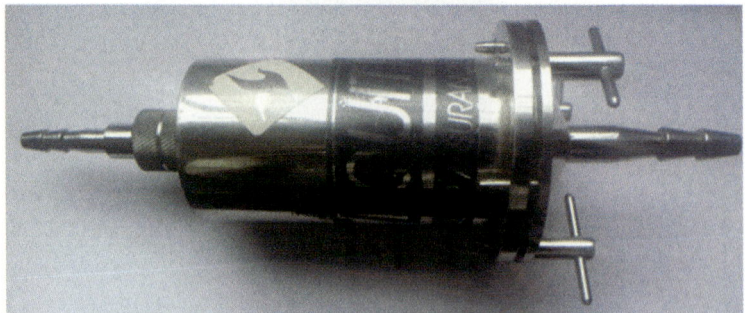

Fig. 5.9: Seitz filter

Use
1. To separate toxins and bacteriophages from bacteria.
2. To obtain bacterial free filtrates of clinical samples for virus isolation.
3. Sterilization of hydatid fluid.

Use the asbestos dics with pore size—0.22 µm
- These filters also remove the mycoplasma
- These filters do not remove viruses.

FAQs
Other types of filter used:
1. Chamberland filters
2. Sintered glass filters
3. Membrane filters—cellulose acetate membrane with pore size of 0.22 µm is used.

Candle Jar (Fig. 5.10)

Use: To provide 3–5% CO_2 which stimulates the growth of *Neisseria* spp, *Haemophilus* spp and *Streptococcus pneumoniae and Brucella abortus*.

Note: It is NOT the method of maintaining anaerobiasis.

Fig. 5.10: Candle jar

McIntosh and Filde's Anaerobic Jar (Fig. 5.11)

It consists of metal or glass jar with a metal lid, attached with a screw, pressure gauge and two openings (inlet and outlet).

Principle: Aluminium pellets coated with palladium are used which removes the residual oxygen by combining it with H_2 to form water.

Use: For cultivation of anaerobic bacteria.

FAQs

Indicators:

i. Biological indicator—plates inoculated with the *Pseudomonas aeruginosa*.

ii. Chemical indicator is reduced methylene blue—it is normally blue in presence of oxygen but become colorless in anaerobic conditions.

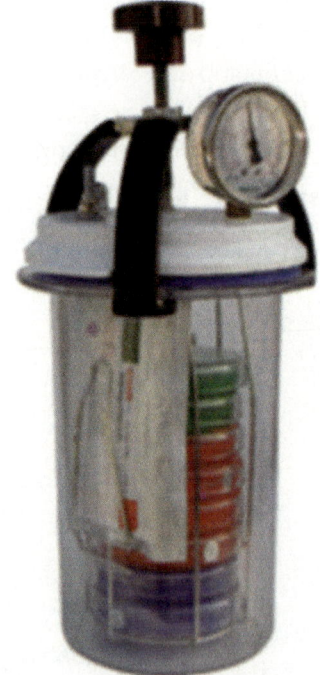

Fig. 5.11: McIntosh and Filde's jar

Tuberculin Syringe (Fig. 5.12)

Use

1. *Used in Mantoux test*: 0.1 ml PPD or 5 TU injected intradermally in the flexor aspect of the arm. Induration of ≥10 mm noted after 48–72 hrs is considered as positive.

2. *Interpretation*: Positive test—indicates only present or past exposure but do not confirm the presence of active infection.

FAQs

1. *False positive Mantoux test*:
   i. BCG vaccination
   ii. NTM (non-tuberculous mycobacteria infection).

2. *False negative Mantoux test*: Early or advanced TB, miliary TB and decreased immunity in HIV positive patients.

3. *Recommended strain for PPD preparation*: PPD-RT-23 with tween 80.

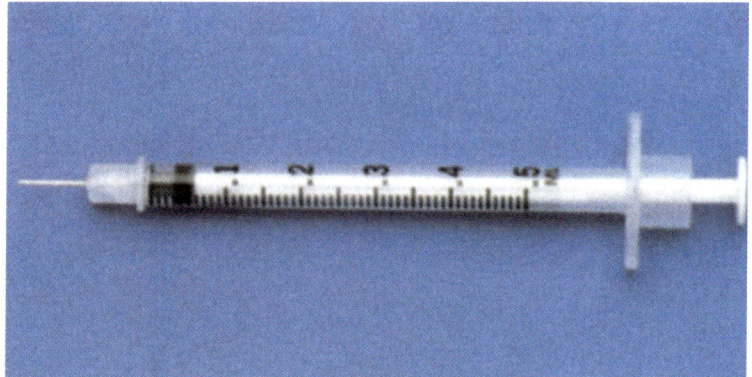

Fig. 5.12: Tuberculin syringe

Water Bath (Fig. 5.13)

Works at temperature

1. 56°C for 1 hr—for serum or heat labile body fluids.
2. 60°C for 1 hr—for bacterial vaccines.

Fig. 5.13: Water bath

Inspissator (Fig. 5.14)

Working temperature
1. 80–85°C for 30 minutes for 3 consecutive days. The first exposure destroys the vegetative forms of bacteria and remaining spores germinate in between intervals and thus killed on subsequent heating.
2. *Use*: Sterilization of serum or egg-based media (Lowenstein-Jensen medium and Loeffler's serum slope).

Fig. 5.14: Inspissator

ELISA Plate (Fig. 5.15)

Enzyme-linked immunosorbent assay (ELISA) is used mainly in immunology to detect the presence of an antibody or an antigen in a sample.

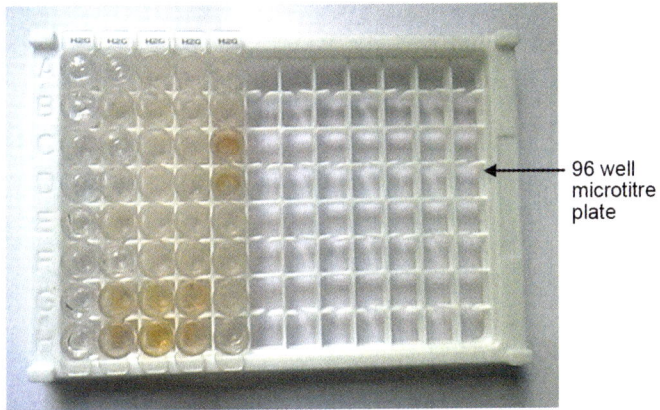

96 well microtitre plate

Fig. 5.15: ELISA plate

Principle: An enzyme conjugated with an antibody reacts with colorless substrate to generate a colored product. The intensity of color/optical density can be measured at 450 nm. This intensity of color is an indication of the amount of antigen or antibody.

Principal components involved in ELISA

1. Solid phase, e.g. polystyrene, polyvinyl or polycarbonate **96 well microplate.**
2. Enzyme conjugate
3. Substrate

Enzymes used

- Alkaline phosphatase
- Horse radish peroxidase
- β-galactosidase

Substrates

- Paranitrophenyl phosphate (pNPP)
- O-phenylene diamine dihydrochloride (OPD)
- Tetramethyl benzidine

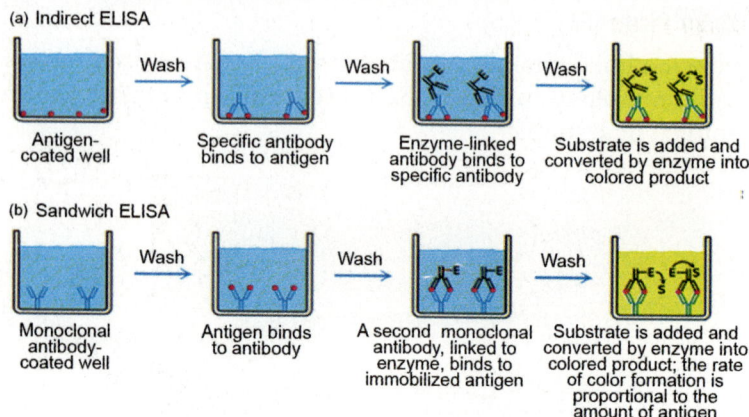

Fig. 5.16: Indiret and sandwich ELISA

Uses

1. Detection of rotavirus **antigen** in feces.
2. Detection of anti-HIV **antibody** in serum.
3. Dengue NS1-antigen and antibody detection.

FAQs

Types of elisa

Type of ELISA generation	Characteristic feature
1st generation ELISA	Viral lysates as antigen
2nd generation ELISA	Recombinant proteins as antigen
3rd generation ELISA	Synthetic peptides as antigen
4th generation ELISA	Both antigen and antibody detection

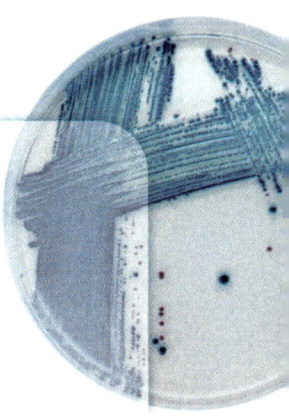

Chapter 6

Culture Media

Culture media are solutions containing all of the nutrients and necessary physical growth parameters necessary for microbial growth.

Bacteria that cannot be grown in artificial culture medium. Examples:
- *Mycobacterium leprae*
- *Rickettsia*
- *Chlamydia*
- *Treponema pallidum*

Bacterial culture media can be classified on the basis of consistency, composition and purpose.

Classification Based on Consistency

1. *Solid medium*: Solid medium is media containing **agar** (at a concentration of **2.0%** or some other, mostly inert solidifying agent).
2. *Semisolid media*: They are prepared with **agar** at concentrations of **0.5%** or less and are useful for the cultivation of microaerophilic bacteria or for determination of bacterial motility.
3. *Liquid (broth) medium*: These media contains specific amounts of nutrients but don't have trace of gelling agents such as gelatin or agar. Broth medium serves various purposes such as propagation of large number of organisms, fermentation studies, and various other tests. *Example*: **Sugar fermentation tests, MR-VR broth**.

Classification Based on the Basis of Composition

1. *Synthetic or chemically defined medium*: A chemically defined medium is one prepared from purified ingredients and, therefore, whose exact composition is known.

2. *Non-synthetic or chemically undefined medium*: Non-synthetic medium contains at least one component that is neither purified nor completely characterized nor even completely consistent from batch to batch. Often these are partially digested proteins from various organism sources. **Nutrient broth**, for example, is derived from cultures of yeasts.

On the basis of their application or function, they are classified as follows:

1. *General purpose media/basic media*: Basal media are basically simple media that supports most non-fastidious bacteria. *Examples*:
   - Peptone water
   - Nutrient broth
   - Nutrient agar

2. *Enriched medium*: Addititon of extra nutrients in the form of blood, serum, egg yolk, etc. to basal medium makes them enriched media. This is done to grow nutritionally exacting (fastidious) bacteria. It includes the following media:
   - Blood agar
   - Chocolate agar
   - Loeffler's serum slope

3. *Selective and enrichment media*: These are designed to inhibit unwanted commensal or contaminating bacteria and help to recover pathogen from a mixture of bacteria. While selective media are agar based, enrichment media are liquid in consistency. Both these media serve the same purpose. Any agar media can be made selective by addition of certain inhibitory agents that don't affect the pathogen. Various approaches to make a medium selective include addition of antibiotics, dyes, chemicals, alteration of pH or a combination of these.

Examples of selective media include:
- Thayer-Martin agar used to recover *N. gonorrhoeae* contains vancomycin, colistin and nystatin.
- Mannitol salt agar and salt milk agar used to recover *S. aureus* contain 10% NaCl.
- Potassium tellurite media containing 0.04% potassium tellurite are used to recover *C. diphtheriae.*
- McConkey's agar used for Enterobacteriaceae members contains bile salt that inhibits most Gram-positive bacteria.
- Pseudosel agar (cetrimide agar) used to recover *P. aeruginosa* contains cetrimide (antiseptic agent).
- Crystal violet blood agar used to recover *S. pyogenes* contains 0.0002% crystal violet.
- Selective media such as TCBS agar used for isolating *V. cholerae* from fecal specimens have bile salts which inhibit most other bacteria.

4. *Enrichment culture/medium*: Enrichment medium is used to increase the relative concentration of certain microorganisms in the culture prior to plating on solid selective medium. Enrichment media are liquid media that also serves to inhibit commensals in the clinical specimen. *Example*:
- Selenite F broth
- Tetrathionate broth
- Akaline peptone water (APW)

5. *Differential/ indicator medium*: Certain media are designed in such a way that different bacteria can be recognized on the basis of their colony color. Various approaches include incorporation of dyes, metabolic substrates, etc. so that those bacteria that utilize them appear as differently colored colonies. Such media are called differential media or indicator media. Differential media allow the growth of more than one micro-organism of interest but with morphologically distinguishable colonies. Examples of differential media include:
- Mannitol salt agar (mannitol fermentation = yellow)
- Blood agar (various kinds of hemolysis, i.e. α, β and γ hemolysis)
- MacConkey agar (lactose fermenters produce pink colonies whereas non-lactose fermenter produces pale or colorless colonies).

- TCBS (*Vibrio cholerae* produces yellow colonies due to fermentation of sucrose).

6. *Transport media*: Clinical specimens must be transported to the laboratory immediately after collection to prevent overgrowth of contaminating organisms or commensals. Such media prevent drying (desiccation) of specimen and inhibit overgrowth of unwanted bacteria. Examples:
   - Stuart's and Amie's medium (semi-solid in consistency). Addition of charcoal serves to neutralize inhibitory factors.
   - Cary Blair medium and Venkatraman Ramakrishnan (VR) medium are used to transport feces from suspected cholera patients.
   - Sach's buffered glycerol saline is used to transport feces from patients suspected to be suffering from bacillary dysentery.
   - Pike's medium is used to transport *Streptococci* from throat specimens.

7. *Anaerobic media*: Anaerobic bacteria need special media for growth because they need low oxygen content, reduced oxidation—reduction potential and extra nutrients.
   - Media for anaerobes may have to be supplemented with nutrients like hemin and vitamin K.
   - Addition of 1% glucose, 0.1% thioglycollate, 0.1% ascorbic acid, 0.05% cysteine or red hot iron filings can render a medium reduced.
   - Before use the medium must be boiled in water bath to expel any dissolved oxygen and then sealed with sterile liquid paraffin.
   - *Examples*:
     - Robertson cooked meat broth commonly used to grow *Clostridium* sp, contains a 2.5 cm column of bullock heart meat and 15 ml of nutrient broth.
     - Thioglycollate broth contains sodium thioglycollate, glucose, cystine, yeast extract and casein hydrolysate.
     - Methylene blue or resazurin is an oxidation—reduction potential indicator that is incorporated in the medium. Under reduced condition, methylene blue is colorless.

Chapter 7

Individual Media

Peptone Water (Fig. 7.1a)

Peptone water is a simple basal liquid medium, used as a growth medium and as a base for carbohydrate fermentation media.

Composition

1. Peptic digest of animal tissue (peptone)
2. Sodium chloride
3. Final pH – 7.2

Note: If desired add required carbohydrate for checking fermentation pattern with added 1% phenol red solution.

Use

1. Peptone water is particularly suitable as a substrate in the study of indole production.

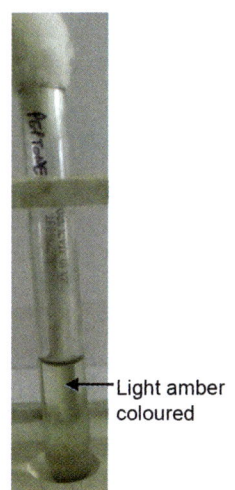

Light amber coloured

Fig. 7.1a: Peptone water

2. Peptone water is also utilized as a base for carbohydrate fermentation studies with the addition of sugar (1%) and indicators such as bromocresol purple, phenol red or bromothymol blue.
3. Peptone water with pH adjusted to 8.4 (alkaline peptone water) is suitable for the cultivation and enrichment of *Vibrio* species.

Appearance: Light **amber-colored** clear solution without any precipitate

Sterilization: Autoclave 121°C at 15 lbs for 15 minutes.

Nutrient Broth (Fig. 7.1b)

Nutrient broth is used for the general cultivation of less fastidious microorganisms, can be enriched with blood or other biological fluids.

Composition: Meat extract + Peptone water
Final pH – 7.4±0.2

Sterilization: Autoclave 121°C at 15 lbs for 15 minutes.

Use
1. Maintaining microorganisms
2. Cultivating fastidious organisms by enriching with serum or blood
3. For study of bacterial growth curve

Appearance: Light **amber-colored** clear solution in tubes.

Classification
It is of 3 types:
1. Meat extract—meat extract and commercial peptone.
2. Digest broth—aqueous extract of lean meat with proteolytic enzyme.
3. Meat infusion—aqueous extract of lean meat and peptone.

Fig. 7.1b: Nutrient broth

Nutrient Agar (Nutrient Broth + 2% Agar)

Nutrient agar is a solid basal medium used for the cultivation of less fastidious microorganisms, can be enriched with blood or other biological fluids (Figs 7.2 to 7.4).

Use
1. For the cultivation and enumeration of bacteria which are not particularly fastidious.
2. Addition of different biological fluids such as horse or sheep blood, serum, egg yolk, etc. makes it suitable for the cultivation of related fastidious organisms.
3. For performing oxidase test, catalase test, slide agglutination test.
4. For pigment demonstration like golden yellow colony of *Staphylococcus aureus.*

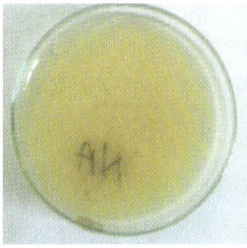
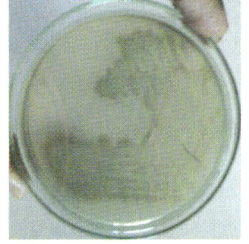
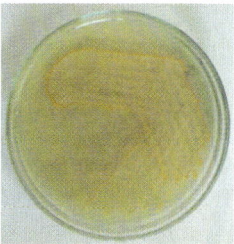

Fig. 7.2: Uninoculated nutrient agar

Fig. 7.3: Greenish pigment of pseudomonas spp

Fig. 7.4: Golden yellow colony – *Staphylococcus aureus*

Appearance: Light **yellow-colored** clear to slightly opalescent in petri plate.

Blood Agar (Fig. 7.5)

Blood agar (BA) is an enriched medium used to culture those bacteria or microbes that do not grow easily. Such bacteria are called "**fastidious**" as they demand a special, enriched nutritional environment as compared to the routine bacteria, such as *Haemophilus influenzae, Streptococcus pneumoniae* and *Neisseria* species.

Composition of blood agar: *(Nutrient agar + 5–10% sheep blood)*

pH—7.2–7.6

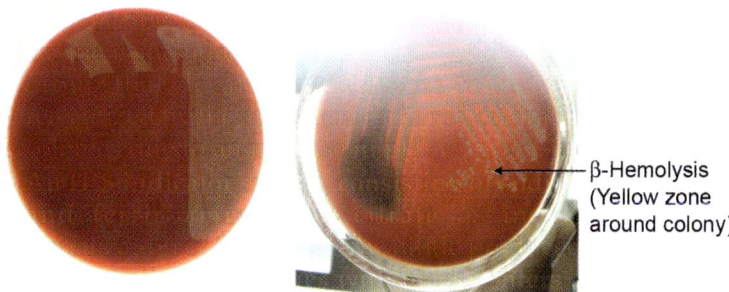

β-Hemolysis (Yellow zone around colony)

Fig. 7.5: Blood agar and beta haemolysis on blood agar

Once the nutrient agar has been autoclaved, allow it to cool but not solidify. When the agar has cooled to 45–50°C, add **5% sterile defibrinated blood** that has been warmed to room temperature and mix gently but well.

Uses of blood agar
1. Growth of fastidious organisms like as *Haemophilus influenzae, Streptococcus pneumoniae* and *Neisseria* species**.**
2. To differentiate bacteria based on their hemolytic properties **[β-hemolysis, α-hemolysis and γ-hemolysis (or non-hemolytic)].**

Blood agar can be made selective for certain pathogens by the addition of antibiotics, chemicals or dyes. *Examples include*:
- Crystal violet blood agar to select *Streptococcus pyogenes* from throat swabs.
- Kanamycin or neomycin blood agar to select anaerobes from pus.

Chocolate Agar (Fig. 7.6)

Chocolate agar is the lysed blood agar. The name is itself derived from the fact that red blood cell (RBC) lysis gives the medium a chocolate-brown color. It is used for the isolation of fastidious organisms, such as *Haemophilus influenzae, Neisseria gonorrhoeae, Neisseria meningitidis* when incubated at 35–37°C in a 5% CO_2 atmosphere.

Fig. 7.6: Chocolate agar

The **lysis of RBC** during the heating process releases intracellular **coenzyme nicotinamide adenine dinucleotide (NAD or V factor)** into the agar for utilization by fastidious bacteria (the heating process also inactivates growth inhibitors). **Hemin (factor X) is available from non-hemolyzed as well as hemolyzed blood cells.**

Preparation of chocolate agar: Add sterile blood to molten nutrient agar when it is **75°C hot.** (*Note*: If blood is added when temperature is 55°C, it will be blood agar.)

Modification of chocolate agar: *Thayer-Martin agar*—it is used for the selective isolation of *N. gonorrhoeae* and *N. meningitidis*. Thayer-Martin medium is a chocolate agar supplemented with vancomycin, nystatin and colistin to inhibit the normal flora, including nonpathogenic *Neisseria*.

MacConkey Agar (Fig. 7.7)

MacConkey agar is a **selective and differential culture media** commonly used for the isolation of enteric Gram-negative bacteria. (Gram-negative enteric bacteria can tolerate to bile salt present in the medium because of their bile-resistant outer membrane.) It helps to **differentiate lactose fermenters from non-lactose fermenters** (Figs 7.8 and 7.9).

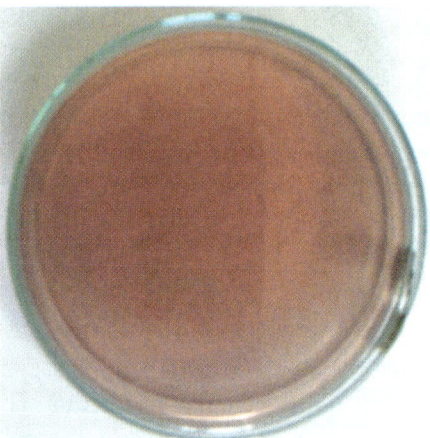

Fig. 7.7: Uninoculated MacConkey agar

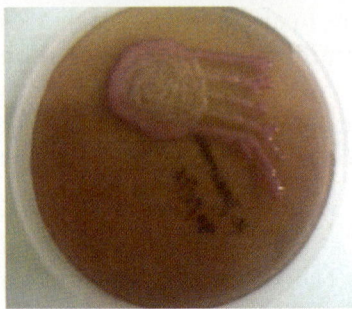
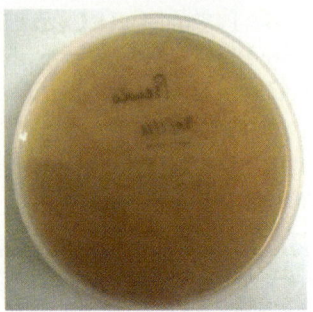

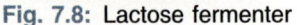

Fig. 7.8: Lactose fermenter   Fig. 7.9: Non-lactose fermenter

Crystal violet and bile salts are incorporated in MacConkey agar to prevent the growth of Gram-positive bacteria and fastidious Gram-negative bacteria, such as *Neisseria* and *Pasteurella*.

Composition of MacConkey agar

1. Enzymatic digest of gelatin, casein and animal tissue: provides nitrogen, vitamins, minerals and amino acids essential for growth (peptone).
2. *Lactose*: Fermentable carbohydrate providing carbon and energy.
3. *Bile salts*: Selective agents and inhibit Gram-positive organisms. Inhibit swarming of *Proteus* spp.
4. *Crystal violet*: To inhibit gram positive bacteria.
5. *Sodium chloride*: Supplies essential electrolytes for transport and osmotic balance.
6. *Neutral red*: pH indicator, which is red in color at pH below 6.8. When lactose is fermented, the pH of the medium decreases, changing the color of neutral red to pink.

Sterilization: Autoclave 121°C at 15 lbs for 15 minutes.

Interpretation

1. **Strongly lactose fermenting bacteria** appear as pink colonies. For example,
   - *Klebsiella spp*: Mucoid lactose fermenter
   - *Escherichia coli*: Flat, dry, pink colonies with a surrounding darker pink area of precipitated bile salts.

2. Gram-negative bacteria that do not ferment lactose appear colorless on the medium called as **non-lactose fermenters**. For example,
   - *Proteus* spp
   - *Shigella* spp
   - *Yersinia* spp
   - *Salmonella* spp

   Other organisms showing colorless colonies are: *Edwardsiella* spp., *Hafnia* spp., *Morganella* spp., *Providencia* spp.
3. *Late lactose fermenter*:
   - *Citrobacter* spp
   - *Shigella sonnei*

FAQs

MacConkey sorbitol agar: EHEC 0157 is non-sorbitol fermenting, producing colorless colonies. Most other *E. coli* strains and other enterobacteria ferment sorbitol. Sorbitol-fermenting organisms produce pink colonies.

Tellurite Blood Agar (Fig. 7.10)

Tellurite blood agar is used for the selective isolation and cultivation of *Corynebacterium* species.

Principle: *C. diphtheriae* reduces potassium tellurite to tellurium and thereby produce gray-black-colored colonies (Fig. 7.11). Throat or nasal swab is directly inoculated and streaked on this agar medium.

Fig. 7.10: Uninoculated tellurite media

Fig. 7.11: Gray-black-colored colonies

Corn starch present in medium neutralizes the toxic metabolites. Haemoglobin and vitamin growth supplement stimulate good growth of *Corynebacterium*.

Loeffler's Serum Slope (Fig. 7.12)

Enriched media for *Corynebacterium diphtheriae*.

Composition

1. Heart muscle, infusion from peptic digest of animal tissue
2. Sodium chloride
3. Dextrose
4. Bovine serum
   Final pH (at 25°C)—7.6 ±0.2

Sterilization: Coagulate and sterilize by **inspissation** for 15 minutes at 80 to 90°C or steaming at 100°C for 10–15 minutes.

Fig. 7.12: Loeffler serum slope

Lowenstein-Jensen (LJ) Media (Fig. 7.13)

Composition of LJ medium

1. Potato flour (potato starch)—as a source of vitamins.
2. L-asparagine—as a source of nitrogen
3. Egg—enrich the medium
4. **Glycerol**—encourages the earliest possible growth of **human mycobacteria** (Fig. 7.14).
5. Monopotassium phosphate—enhance organism growth and act as buffer
6. Magnesium citrate
7. Magnesium sulfate—enhance organism growth and act as buffer.
8. Malachite green—prevents growth of the majority of contaminants and provides **green color** to the medium.

For cultivation of *M. bovis*, glycerol is omitted and sodium pyruvate is added.

When heated, the egg albumin coagulates, thus providing a solid surface for inoculation.

Fig. 7.13: Uninoculated LJ media

Fig. 7.14: Growth of *Mycobacterium tuberculosis* on LJ media

Fig. 7.15: NTM growth on LJ media

Uses of LJ medium

1. Selective media for *Mycobacterial* spp.
2. It is used for testing antibiotic susceptibility of *Mycobacterial* isolates.
3. It is also used for differentiating species of *Mycobacterium* (by colony morphology, growth rate, biochemical characteristics and microscopy).

Colony morphology on LJ medium: *M. tuberculosis*—typical non-pigmented, rough, tough and buff (yellow) colonies.

Cultures should be read within 5–7 days after inoculation and once a week thereafter for up to 8 weeks.

Sterilization: Inspissator at 85°C for 45 minutes.

Alkaline Peptone Water (Fig. 7.16)

Enrichment medium for *Vibrio* species.

Composition

1. Peptic digest of animal tissue (peptone)
2. *Sodium chloride*: Final pH—**8.4 ± 0.2**

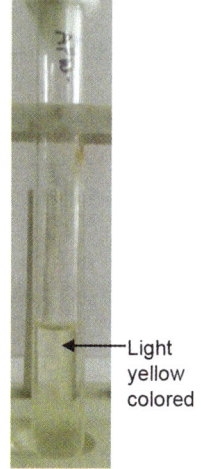

Light yellow colored

Fig. 7.16: Alkaline peptone water

Sterilization: Autoclave at 121°C × 15 lbs × 15 minutes.

Use: Stool samples containing small numbers of *Vibrio* spp should be inoculated into an enrichment medium prior to plating on to a selective medium, such as TCBS agar.

Appearance: Light **yellow-colored** clear solution without any precipitate.

Glucose Broth (Fig. 7.17)

Glucose broth is used for study of glucose (dextrose) fermentation where a pH indicator is not desired.

Composition
1. Casein enzymic hydrolysate
2. Glucose
3. *Sodium chloride*: Final pH (at 25°C) 7.3 ± 0.2

Sterilize by filtration.

Uses
1. To study glucose fermentation using only pure **1% glucose** as the source of carbohydrate.
2. As blood culture media.

Fig. 7.17: Glucose broth

Selenite Broth (Selenite F Broth)

Selenite broth is recommended as enrichment media for the **isolation of** *Salmonella* spp from feces and urine.

It is NOT autoclaved as excessive heating is detrimental.

Uses
1. For detecting *Salmonella* in the non-acute stages of illness when organisms occur in the feces in low numbers.
2. For epidemiological studies to enhance the detection of low number of organisms from asymptomatic or convalescent patients.

Subcultured on selective media like XLD or DCA after 6 to 8 hours.

Xylose Lysine Deoxycholate (XLD) Agar

XLD agar is a selective medium for the isolation of *Salmonella and Shigella* spp from clinical specimens like stool and food samples (Figs 7.18 to 7.20).

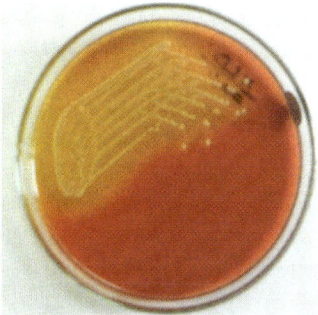

Fig. 7.18: *Shigella flexneri*

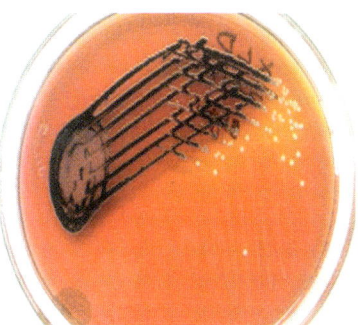

Fig. 7.19: *Salmonella typhi*

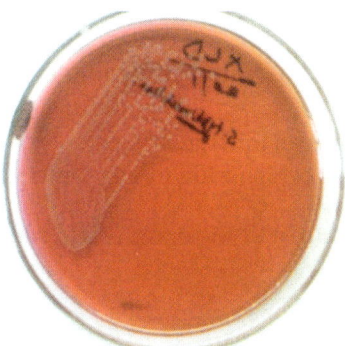

Fig. 7.20: *Escherichia coli*

Principle of XLD agar: XLD agar is both a selective and differential medium. It contains:

1. Xylose is fermented by practically all enteries except for the *Shigella* and this property enables the differentiation of Shigella species.

2. Lysine is included to enable the identification of *Salmonella* group that decarboxylate lysine to cadaverine can be recognized by the appearance of a red coloration around the colonies due to an increase in pH.

3. **An H₂S indicator system, consisting of sodium thiosulfate and ferric ammonium citrate**, is included for the visualization of the hydrogen sulfide produced, resulting in the formation of **colonies with black centers**.

Note: Do not autoclave.

Typical colonial morphology on XLD agar are as follows:
- *Salmonella typhi*: Red colonies, black centers
- *Salmonella choleraesuis*: Red colonies
- *Shigella sonnei*: Red colonies
- *Shigella flexneri*: Red colonies
- *Escherichia coli*: Yellow colonies
- *Proteus vulgaris*: Yellow colonies
- *Pseudomonas aeruginosa*: Pink colonies.

Thioglycollate Broth

Thioglycollate broth (fluid thioglycollate medium) is a medium designed to test the aerotolerance of bacteria.

It contains sodium thioglycollate, thioglycollic acid, L-cysteine, methylene blue, and 0.05% agar to reduce the oxygen to water.

Methylene blue is an indicator that is colorless in an anaerobic environment and greenish-blue in the presence of oxygen.

Robertson Cooked Meat Medium (RCM Medium) (Fig. 7.21)

Uses
1. For cultivation of anaerobes, especially pathogenic *Clostridia* spp.
2. As a maintenance medium for stock cultures.

Composition
1. Beef heart—provide amino acids, contains glutathione, a catalyst of reducing reaction and sulfhydryl groups that lower redox potential.
2. Proteose peptone
3. Dextrose—allows rapid and heavy growth of anaerobic bacteria
4. Sodium chloride

Final pH (at 25°C): 7.2 ± 0.

Sterilization: Autoclave 121°C at 15 lbs for 15 minutes.

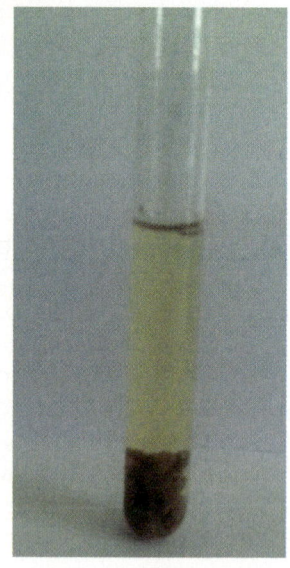

Fig. 7.21: RCM media

Interpretation
1. **Saccharolytic** bacteria **decompose sugars** to form butyric and acetic acids and alcohols. The **meat** in Robertson's medium is **reddened** and gas is produced.
2. **Proteolytic** species attack the amino acids. **Meat** in Robertson's medium is **blackened** and decomposed by *Clostridium* species, giving the culture a foul odor.

Brain Heart Infusion Broth

Enriched liquid medium for **blood culture** (Fig. 7.22).

Composition
1. **Calf brain, infusion**—sources of carbon, nitrogen, essential growth factors, amino acids and vitamins.
2. **Beef heart, infusion**—sources of carbon, nitrogen, essential growth factors, amino acids and vitamins.
3. Proteose peptone
4. Dextrose—source of energy.
5. Sodium chloride—maintains the osmotic equilibrium of the medium
6. Disodium phosphate—buffering action

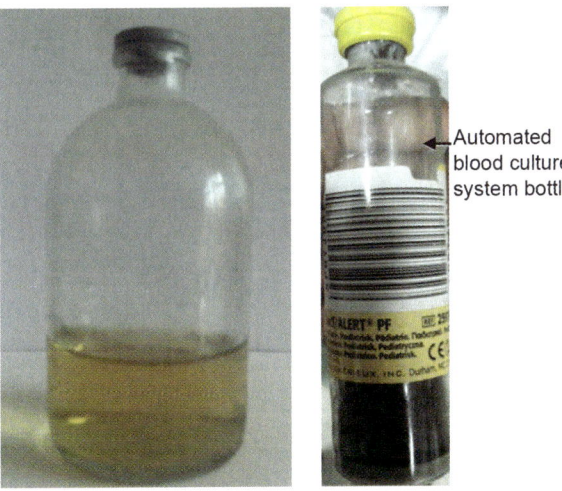

Automated blood culture system bottle

Fig. 7.22: Blood culture bottle

Final pH (at 25°C): 7.4 ±0.2.

Uses

1. Brain heart infusion medium is useful for cultivating a wide variety of microorganisms since it is a highly nutritive medium.
2. It is also used to prepare the inoculums for antimicrobial susceptibility testing.
3. Brain heart infusion broth is also the preferred medium for anaerobic bacteria, yeasts and molds.
4. This medium with the addition of 10% defibrinated sheep blood, it is useful for isolation and cultivation of *Histoplasma capsulatum* and other fungi.

Stuart Transport Medium (Modified)

A semi-solid, non-nutritional transport medium for preservation of fastidious, pathogenic organisms *Neisseria* species.

Appearance of prepared medium: Off white colored gel, semi-solid gel.

CLED Agar Media (Fig. 7.23)

CLED agar: CLED (cysteine lactose electrolyte-deficient) agar with or without Andrade indicator is a differential culture medium for use in isolating and enumerating bacteria in urine from the suspected cases of urinary tract infection. It supports the growth of all potential urinary pathogens including Grampositive organism and *Proteus* spp. It inhibits swarming of *Proteus* spp as it is electrolyte

Fig. 7.23: CLED agar media

deficient. Andrade indicator gives pink colony in case of lactose fermenting bacteria.

CHROM agar (Fig. 7.24): The chromogen mix consists of artificial substrates (chromogens), which release differently colored compounds upon degradation by specific enzymes.

Fig. 7.24: CHROM agar media

This permits the differentiation of certain species or the detection of certain groups of organisms with only a minimum of confirmatory tests.

Sabouraud Dextrose Agar (SDA) (Figs 7.25 and 7.26)

A selective medium primarily used for the isolation of dermatophytes, other fungi and yeasts but can also grow filamentous bacteria such as *Nocardia*.

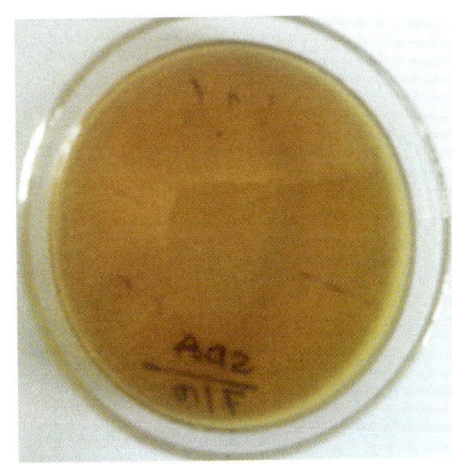

Fig. 7.25: SDA slant   Fig. 7.26: SDA plain

The acidic pH of this medium (**pH about 6.8 in Emmons modification**) inhibits the growth of bacteria but permits the growth of yeasts and most filamentous fungi (Fig. 7.27).

Antibiotics like chloramphenicol and cycloheximide are added to inhibit the growth of bacteria and saprophytic fungi.

Dextrose is the fermentable carbohydrate incorporated in high concentration (**2%**) as a carbon and energy source.

Interpretation
- Yeasts will grow as creamy to white colonies.
- Molds will grow as filamentous colonies of various colors.

Fig. 7.27: Yeast—creamy to white colonies

Typical colony morphology of some fungi in SDA (Fig. 7.28):
1. *Aspergillus flavus*: Yellow-green, powdery and pale yellowish on reverse.
2. *Aspergillus niger*: The initial growth is white, becoming black later on giving "salt and pepper appearance" which results from darkly pigmented conidia borne in large numbers on conidiophores and reverse turning pale yellow.
3. *Aspergillus fumigatus*: Blue-green, powdery and pale yellow on reverse.
4. *Aspergillus nidulans*: Greenish-blue with whitish edge, yellow to brownish on reverse.

Fig. 7.28: *Aspergillus* spp

Limitations
1. It does not promote conidiation of filamentous fungi.
2. Antimicrobial agents added into a medium to inhibit bacteria may also inhibit certain pathogenic fungi.

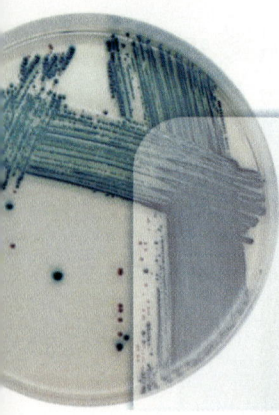

Chapter 8

Tests for Bacterial Identification

Catalase Test (Fig. 8.1)

Principle: Catalase is an enzyme that decompose hydrogen peroxide into water and nascent oxygen.

It is used to differentiate members of the Micrococcaceae (like *Staphylococcus* spp) from members of the Streptococcaceae (*Streptococcus* spp).

Procedure: With an inoculating needle or wooden applicator stick, transfer growth from centre of the colony to the surface of a glass slide and add one drop of **3% hydrogen peroxide** and observe for effervescence (bubble) formation.

Interpretation: The rapid and sustained appearance of bubble or effervescence constitute positive result.

Don't take growth from blood agar and don't use iron wireloop to avoid false positive reaction.

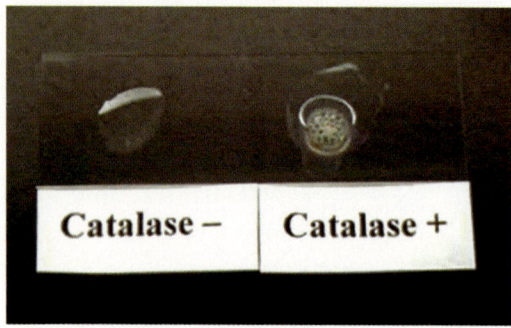

Fig. 8.1: Catalase test

Coagulase Test

Coagulase test is used to differentiate *Staphylococcus aureus* (positive) from coagulase negative *Staphylococcus* (CONS). Coagulase is an enzyme produced by *S. aureus* that converts (soluble) fibrinogen in plasma to (insoluble) fibrin. *Staphylococcus aureus* produces two forms of coagulase, bound and free coagulase:

- Slide coagulase test is done to detect bound coagulase or clumping factor.
- Tube coagulase test is done to detect free coagulase.

Slide Coagulase Test (Fig. 8.2)

Detects clumping factor (formerly referred as cell-bound coagulase).

- Clumping factor directly converts fibrinogen to fibrin causing agglutination.
- Heavy suspension of organism is made on glass slide and mixed with drop of plasma.
- Agglutination (clumping) **within 10 seconds** indicates a positive test:
  - Indicates *Staphylococcus aureus*
  - Some species of coagulase negative *Staphylococcus* can be positive.
- Negative results should be confirmed by tube coagulase test.

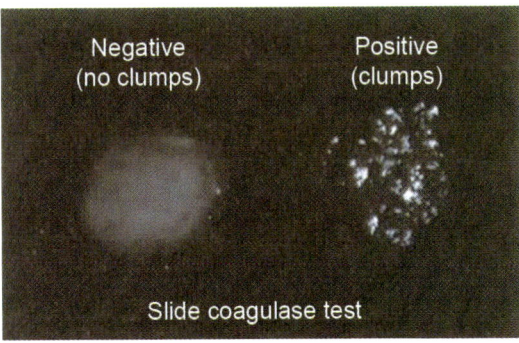

Fig. 8.2: Slide coagulase test

Tube Coagulase Test (Fig. 8.3)

Detects staphylocoagulase which reacts with **coagulase-reacting factor (CRF)**. CRF is a thrombin-like molecule. Staphylocoagulase and CRF combine to indirectly convert fibrinogen to fibrin.

Procedure: A suspension of organism is incubated with plasma at 37°C. Clot formation within 4 hours indicates a positive test.

• Positive test indicates *Staphylococcus aureus*
• Some species of coagulase negative *Staphylococcus* can be positive

Negative tubes should be held overnight at room temp.

Tube coagulase test procedure

1. Prepare a **1-in-6 dilution** of the plasma in saline (0.85% NaCl) and place 1 ml volumes of the diluted plasma in small tubes.
2. Emulsify several isolated colonies of test organism in 1 ml of diluted rabbit plasma to give a milky suspension.
3. Incubate tube at 35°C in ambient air or in water bath for 4 hours.
4. Examine at 1, 2 and 4 hour for clot formation by tilting the tube through 90°. **(Clots may liquefy after their formation.)**

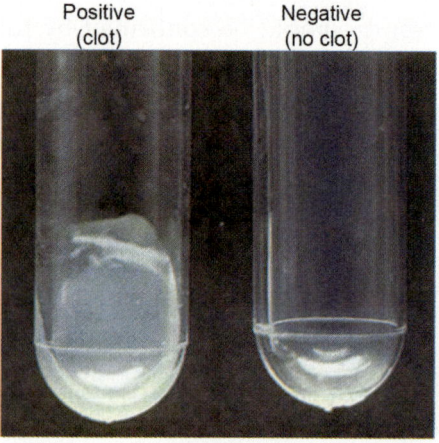

Positive (clot)   Negative (no clot)

Fig. 8.3: Tube coagulase test coagulum formation

Observation: Read as positive any degree of clot formation. Often the plasma is converted into stiff gel that remains in place when the tube is tilted or inverted, but sometimes clots are seen floating in the fluid.

Oxidase Test (Fig. 8.4)

The oxidase test is used to identify bacteria that produce cytochrome-c oxidase, an enzyme of the bacterial electron transport chain.

Principle: When present, the cytochrome-c oxidase oxidizes the **reagent (tetramethyl-p-phenylenediamine dihydrochloride)** to (indophenols), purple color end product. When the enzyme is not present, the reagent remains reduced and is colorless.

Expected results of oxidase test

- *Positive*: Development of dark purple color **(indophenols) within 10 seconds**. Bacterial genera characterized as oxidase positive include *Neisseria* **and** *Pseudomonas*.
- *Negative*: Absence of color. Genera of the Enterobacteriaceae family are characterized as oxidase negative.

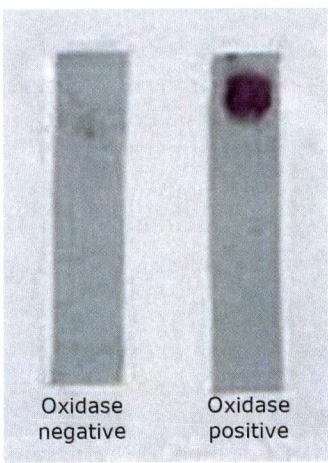

Oxidase negative   Oxidase positive

Fig. 8.4: Oxidase test on strip

Mnemonic for oxidase positive organisms: **PVNCH** (it is just an acronyms inspired by the famous *mnemonic* for urease positive organisms—PUNCH)

- **P:** *Pseudomonas* spp
- **V:** *Vibrio cholerae*
- **N:** *Neisseria* spp
- **C:** *Campylobacter* spp
- **H:** *Helicobacter* spp
- *Aeromonas* spp
- *Alcaligens* spp

Indole Test (Fig. 8.5)

Indole test is used to determine the ability of an organism to **split amino acid tryptophan to** form the compound **indole.**

Principle: Tryptophan is hydrolysed by tryptophanase to produce three possible end products—one of which is indole. Indole production is detected by **Kovac's or Ehrlich's reagent which contains 4 (p)-dimethylaminobenzaldehyde**, this reacts with indole to produce a red-colored compound.

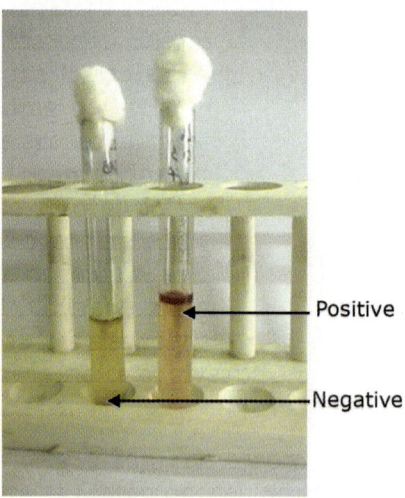

Positive

Negative

Fig. 8.5: Indole test

Indole test is a commonly used biochemical test (e.g. in IMViC test, **SIM test,** etc.). Indole test helps to differentiate Enterobacteriaceae and other genera.

Expected results
- **Positive: Pink-colored ring** after addition of appropriate reagent. Indole positive organisms:
  - *E. coli*
  - *Proteus vulgaris*
  - *Morganella morganii*
  - *Providencia* spp
  - *Vibrio cholerae*
  - *Klebsiella oxytoca*
- **Negative: No color change** even after the addition of appropriate reagent. For example,
  - *Klebsiella pneumoniae*
  - *Salmonella* spp
  - *Proteus mirabilis*

Citrate Utilization Test (Fig. 8.6)

Citrate utilization test is commonly employed as part of a group of tests, the IMViC (indole, methyl red, VP and citrate) tests, that distinguish between members of the Enterobacteriaceae family based on their metabolic byproducts.

Principle: Citrate utlilization test is based on the ability of bacteria to **utilize sodium citrate as its only carbon source and inorganic ($NH_4H_2PO_4$) as the sole fixed nitrogen source**. An alkaline compound formed during citrate utilization reaction raise the pH.

Medium used for citrate utilization test: Simmons citrate agar.

Indicator used: Bromothymol blue.
- *Citrate positive*: Growth will be visible on the slant surface and the medium will be an intense **Prussian blue**.
- *Citrate negative*: Trace or **no growth** will be visible. No color change will occur; the medium will remain the deep forest green color of the uninoculated agar.

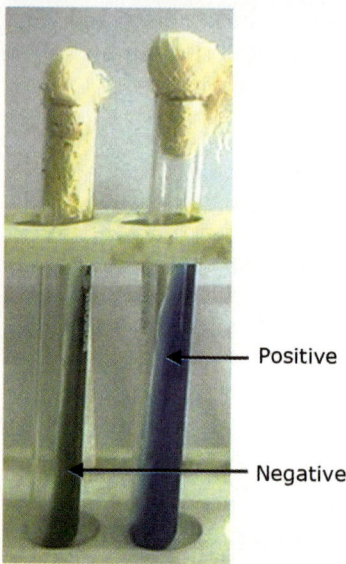

Fig. 8.6: Citrate utilization test

Bacteria which give **positive** citrate utilization test:

- *Klebsiella pneumoniae*
- *Enterobacter* species (minority of strains gives negative result)
- *Citrobacter freundii*
- *Salmonella* other than typhi and paratyphi A
- *Serratia marcescens*
- *Proteus mirabilis* (minority of strains gives negative result)
- *Providencia* spp

Bacteria which give **negative** citrate utilization test:

- *Escherichia coli*
- *Shigella* spp
- *Salmonella typhi*
- *Salmonella paratyphi A*
- *Morganella morganii*
- *Yersinia enterocolitica.*

Urease Test (Fig. 8.7)

Principle: **Urea** is a diamide of carbonic acid. It is **hydrolyzed** by bacteria having urease enzyme, with the release of ammonia and carbon dioxide. The ammonia combines with carbon dioxide and water to form ammonium carbonate which turns the medium alkaline, turning the **indicator phenol red** from its original orange yellow color to bright pink.

Medium used for urease test: Christensen urease media.

Indicator used in urease test: Phenol red.

Organisms that hydrolyze urea rapidly (e.g. *Proteus* spp) may produce positive reactions within 1 or 2 hours; less active species (e.g. *Klebsiella* spp) may require 3 or more days. In routine diagnostic laboratories the urease test result is read within 24 hours.

Name of **urease-positive** organisms:
- *Klebsiella pneumoniae*
- *Proteus* spp
- *Helicobacter pylori*
- *Yersinia* spp

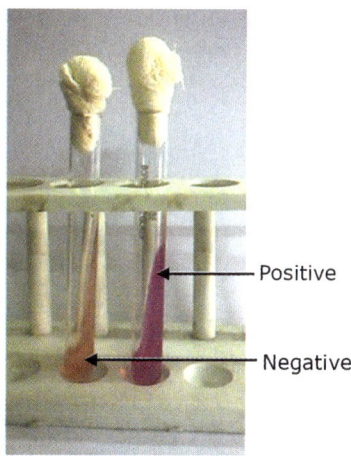

Fig. 8.7: Urease test

- *Brucella* spp
- *Corynebacterium* spp
- *Cryptococcus* spp

Mnemonic to remember **urease-positive** organisms: **PUNCH**
(Similar mneomonic for oxidase positive organism is PVNCH)

- **P:** *Proteus*
- **U:** *Ureaplasma*
- **N:** *Nocardia*
- **C:** *Cryptococcus neoformans/Corynebacterium* spp
- **H:** *Helicobacter pylori*

Urease-negative organisms: *E. coli, Enterobacter aerogens, Salmonella* spp.

Phenylalanine Deaminase Test (Fig. 8.8)

Principle: Phenylalanine deaminase medium tests the ability
of an organism to produce the **enzyme deaminase**. This enzyme
removes the amine group from the amino acid phenylalanine
and releases the amine group as free ammonia. As a result of
this reaction, phenylpyruvic acid is also produced.

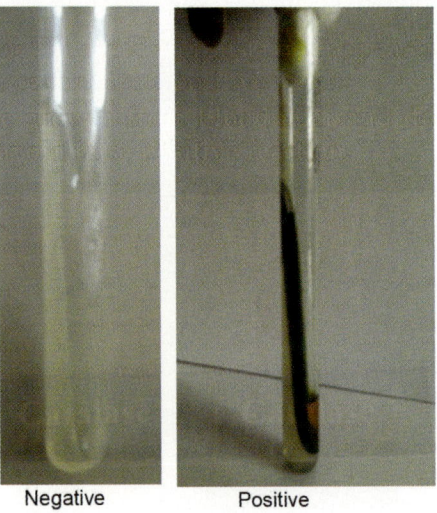

Negative   Positive

Fig. 8.8: PPA test

Phenylalanine agar, also known as phenylalanine deaminase medium, contains nutrients and L-phenylalanine. It is used to differentiate members of the genera *Proteus, Morganella* (which were originally classified under the genus *Proteus*), and *Providencia* from other Enterobacteriaceae.

After incubation, **10% ferric chloride** is added to the media; if phenylpyruvic acid was produced, it will react with the ferric chloride and turn dark green. If the medium remains a straw color, the organism is negative for phenylalanine deaminase production.

Phenylalaninc deaminase test **positive** organisms are:
• *Proteus* spp
• *Morganella* spp
• *Providencia* spp

Phenylalanine deaminase test **negative** organisms are:
• *E. coli*
• *Klebsiella* spp
• *Salmonella* spp

Methyl Red (MR) Test (Fig. 8.9)

Methyl red (MR) test determines whether the microbe performs mixed acids fermentation when supplied glucose.

Mixed acid fermentation is one of the two broad patterns, 2-3-butanediol fermentation being another. In mixed acid fermentation, three acids (acetic, lactic and succinic) are formed in significant amounts. The mixed acid pathway gives of acidic products (mainly lactic and acetic acid).

These large amounts of acid result significant decrease in the **pH of the medium below 4.4**. This is visualized by using pH indicator, **methyl red (p-dimethylaminoazobenzene-O-carboxylic acid),** which is yellow above pH 5.1 and red at pH 4.4.

The pH at which methyl red detects acid is considerably lower than the pH for other indicators used in bacteriologic culture media. Thus, to produce a color change, the test organism must produce large quantities of acid from carbohydrate substrate being used.

MR positive: When the culture medium turns **red** after addition of methyl red, because of a pH at or below 4.4 from the fermentation of glucose.

MR negative: When the culture medium remains **yellow**, which occurs when less acid is produced (pH is higher) from the fermentation of glucose.

MR-VP broth (glucose phosphate peptone broth) is used for both MR test and VP test. Only the addition of reagent differs, and both tests are carried out consecutively.

Expected results

- A **positive reaction** is indicated, if the color of the medium changes to red within a few minutes. MR test positive organisms are:
  - *Escherichia coli*
  - *Citrobacter* spp
  - *Salmonella* spp
- **MR test negative:**
  - *Enterobacter* spp
  - *Klebsiella* spp
  - *Hafnia* spp
  - *Serratia* spp

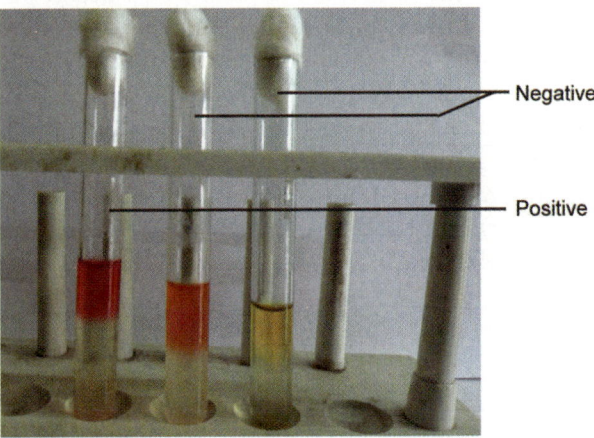

Fig. 8.9: Methyl red test

Voges-Proskauer Test (Fig. 8.10)

Voges-Proskauer is a double eponym, named after two microbiologists working at the beginning of the 20th century.

Principle: Pyruvic acid, the pivotal compound in the fermentative degradation of glucose, is further metabolized by a pathway that results in the **production of acetoin (acetyl methyl carbinol),** a neutral-reacting end product, a product of the **butylenes glycol pathway.**

In the presence of atmospheric oxygen and 40% potassium hydroxide, acetoin is converted to diacetyl, and alpha-naphthol serves as a catalyst to bring out a red complex.

Media and reagents

- *Media*: The medium is **MR/VP broth.**
- *Reagents*:
  1. Alpha-naphthol 5%
  2. Potassium hydroxide, 40%, oxidizing agent

Add 0.6 mL of 5% alpha naphthol, followed by 0.2 mL of 40% KOH to the MR/VP broth with bacterial suspension. Shake the tube gently to expose the medium to atmospheric oxygen and allow the tube to remain undisturbed for 10 to 15 minutes.

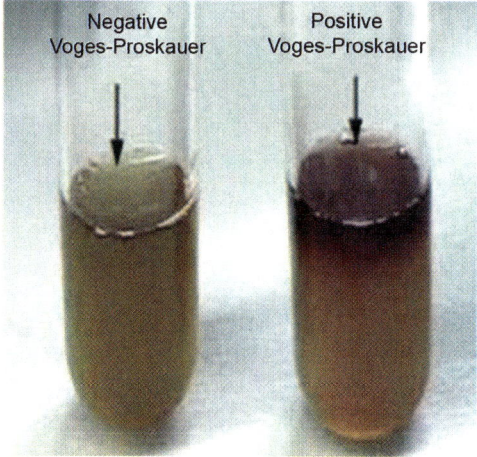

Fig. 8.10: VP test

Results and interpretation

- A **positive test** is represented by the development of a red color 15 minutes or more after the addition of the reagents indicating the presence of diacetyl, the oxidation product of acetoin. **VP test positive organisms** are:
  - *Klebsiella* spp
  - *Enterobacter* spp
  - *Hafnia* spp
  - *Serratia* spp

- The test should not be read after standing for over 1 hour because negative Voges-Proskauer cultures may produce a copper-like color, potentially resulting in a false positive interpretation. **VP test negative organisms** are:
  - *E. coli*
  - *Citrobacter* spp
  - *Salmonella* spp

Triple Sugar Iron (TSI) Test (Fig. 8.11)

TSI tube contains butt (poorly oxygenated area on the bottom) and slant (angled well-oxygenated area on the top).

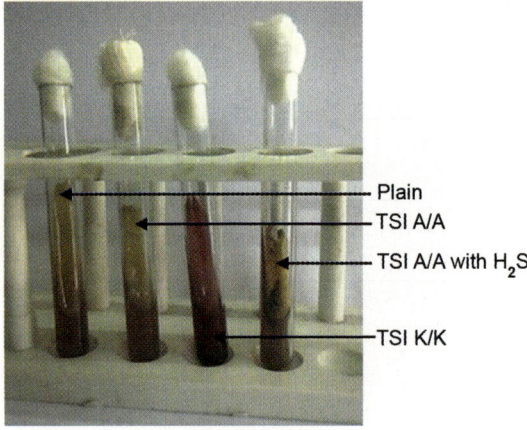

Fig. 8.11: TSI test

Composition of triple sugar iron (TSI) agar

1. Lactose, sucrose and glucose in the concentration of 10:10:1 (i.e. 10 part lactose (1%), 10 part sucrose (1%) and 1 part glucose (0.1%)).
2. **Iron:** *Ferrous sulfate*—indicator of H_2S formation
3. **Phenol red:** *Indicator* of acidification (it is yellow in acidic condition and red under alkaline conditions).
4. It also contains peptone which acts as source of nitrogen.

Remember that whenever peptone is utilized under aerobic condition, ammonia is produced.

TSI is similar to Kligler's iron agar (KIA), except that Kligler's iron agar contains only two carbohydrates: glucose (0.1%) and lactose (1%).

Interpretation

1. *0.1% glucose*: If only glucose is fermented, only enough acid is produced to turn the butt yellow. The slant will remain red.
2. *1.0% lactose/1.0% sucrose*: A large amount of acid turns both butt and slant yellow, thus indicating the ability of the culture to ferment either lactose or sucrose.

Why sucrose is added in TSI?

Addition of sucrose in TSI agar permits earlier detection of coliforms that ferment sucrose more rapidly than lactose. Adding sucrose also aids the identification of certain Gram-negative bacteria that could ferment sucrose but not lactose. **Example:** *Yersenia enterocolitica* and *Vibrio cholerae* which ferments sucrose and not lactose thus giving a TSI reaction as A/A.

Interpretation of triple sugar iron agar test

1. *TSI A/A (acidic slant/acidic butt)*: If lactose (or sucrose) is fermented, a large amount of acid is produced, which turns the phenol red indicator yellow both in butt and in the slant. Some organisms generate gases, which produce bubbles/ cracks on the medium.
2. *TSI K/A(alkaline slant/acidic butt)*: If lactose is not fermented but the small amount of glucose is, the oxygen deficient butt will be yellow (remember that butt comparatively have more glucose compared to slant, i.e. more media more glucose), but on the slant the acid (less acid as media in slant is very

less) will be oxidized to carbon dioxide and water by the organism and the slant will be red (alkaline or neutral pH).

3. *TSI K/K (alkaline slant/no change in butt)*: If neither lactose/ sucrose nor glucose is fermented, both the butt and the slant will be red. The slant can become a deeper red-purple (more alkaline) as a result of production of ammonia from the oxidative deamination of amino acids (remember peptone is a major constitutent of TSI agar).

4. If H_2S is produced, the black color of ferrous sulfide is seen.

Some examples of triple sugar iron (TSI) agar reactions

Name of the organisms	Slant	Butt	Gas	H_2S
Escherichia, Klebsiella, Enterobacter	Acid (A)	Acid (A)	Pos (+)	Neg (−)
Shigella, Serratia	Alkaline (K)	Acid (A)	Neg (−)	Neg (−)
Salmonella, Proteus	Alkaline (K)	Acid (A)	Pos (+)	Pos (+)
Pseudomonas	Alkaline (K)	Alkaline(K)	Neg (−)	Neg (−)

Carbohydrate Fermentation Test (Fig. 8.12)

Principle: It detects the ability of an organism to ferment a specific carbohydrate (sugar) incorporated in a medium producing acid with or without gas. Glucose, lactose, sucrose and mannitol are widely used for sugar fermentation.

Medium: Carbohydrate fermentation broth containing andrade indicator and 1% sugar.

Expected result

Original color: Colorless or slight yellow.

1. *Acid production*: Changes the medium into pink. The organism ferments the given carbohydrate and produce organic acids thereby reducing the pH of the medium into acidic.

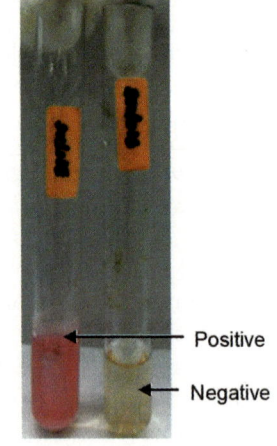

Positive

Negative

Fig. 8.12: Carbohydrate fermentation test

2. *Acid and gas production*: Changes the medium into pink color. The organism ferments the given carbohydrate and produces organic acids and gas. Gas production can be detected by the presence of small bubbles in the inverted **Durham tubes.**

3. *Absence of fermentation*: The broth retains the slight yellow color.

Nitrate Reduction Test (Fig. 8.13)

Principle: Nitrate reduction test is used for differentiation of members of **Enterobacteriaceae** on the basis of their ability to produce nitrate reductase enzyme that hydrolyze nitrate (NO_3^-) to nitrite (NO_2^-) which may then again be degraded to various nitrogen products like nitrogen oxide, nitrous oxide and ammonia (NH_3) depending on the enzyme system of the organism and the atmosphere in which it is growing.

Reagent used in test: Sulfanilic acid and α-naphthylamine.

If the organism has reduced nitrate to nitrite, the nitrites in the medium will form nitrous acid. When sulfanilic acid is added, it will react with the nitrous acid to produce diazotized **sulfanilic acid.** This reacts with the **α-naphthylamine** to form a red-colored compound. Therefore, if the medium **turns red**

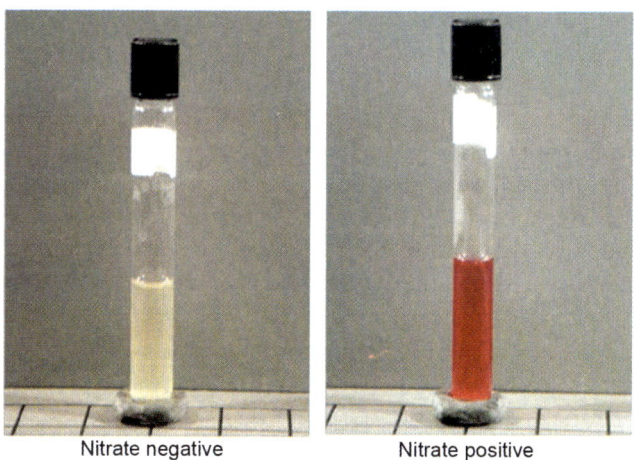

Nitrate negative   Nitrate positive

Fig. 8.13: Nitrate reduction test

after the addition of the nitrate reagents, it is considered a **positive** *result for nitrate reduction. For example,*

- *E. coli*
- *Klebsiella* spp
- *Enterobacter* spp
- *Salmonella* spp

If the medium does not turn red after the addition of the reagents, it can mean that the organism was unable to reduce the nitrate, or the organism was able to denitrify the nitrate or nitrite to produce ammonia or molecular nitrogen.

Therefore, **another step** is needed in the test. Add a small amount of **powdered zinc**. If the tube turns red after the addition of the zinc, it means that unreduced nitrate was present. **Therefore, a red color on the second step is a negative result.**

If the medium does not turn red after the addition of the zinc powder, then the result is called a positive complete.

If no red color forms, there was no nitrate to reduce. Since there was no nitrite present in the medium, either, that means that denitrification took place and ammonia and molecular nitrogen were formed.

Decarboxylase Test (Fig. 8.14)

Decarboxylases are a group of substrate specific enzymes that are capable of reacting with the carboxyl (COOH) portion of amino acids, forming alkaline-reacting amines and byproduct carbon dioxide. Increased pH of the medium is detected by color change of the pH **indicators bromcresol purple and cresol red.**

Bromocresol purple turns purple at an alkaline pH and turns yellow at an acidic pH.

Each decarboxylase enzyme is specific for an amino acid. Lysine, ornithine and arginine are the three amino acids routinely tested in the **identification of Enterobacteriaceae.**

The **specific amine products are:**

1. Decarboxylation of lysine forms cadaverine
2. Decarboxylation of ornithine forms putrescine
3. Decarboxylation of arginine forms citrulline

These byproducts are sufficient to raise the pH of the media so that the broth turns purple.

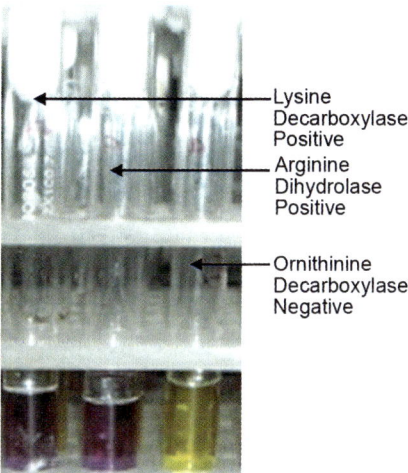

Lysine
Decarboxylase
Positive

Arginine
Dihydrolase
Positive

Ornithinine
Decarboxylase
Negative

Fig. 8.14: Decarboxylase test

Oxidation Fermentation (OF) Test

Oxidation fermentation (OF) test is used to differentiate those organisms that utilize carbohydrates aerobically **(oxidation) such as** *Pseudomonas aeruginosa,* from those that utilize carbohydrates anaerobically **(fermentation) such as members of the Enterobacteriaceae** and those that do not utilize carbohydrates at all **(non-fermenters) such as** *Alcaligenes faecalis.* This test was developed by Hugh and Leifson so the OF medium is known as Hugh Leifson medium.

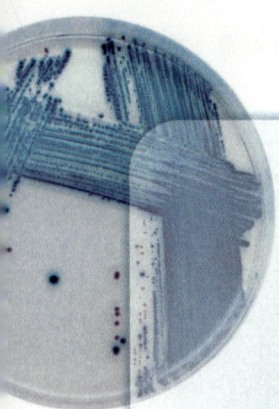

Chapter 9

Serology—Spots

VDRL Tile (Fig. 9.1)

This tile is 5 × 7.5 cm containing 12 concave rings with each ring having an internal diameter of 14 mm.

Type of test: VDRL test also called as non-treponemal micro-flocculation test or **slide flocculation test**.

VDRL test is named after Venereal Disease Research Laboratory, New York, where the test was developed.

Antigen used: Cardiolipin antigen (chemically—diphosphatidyl-glycerol). It is derived from bovine heart.

In India: This antigen is made at Institute of Serology, Kolkata.

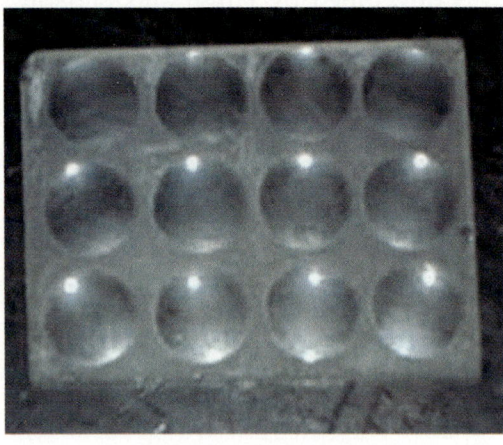

Fig. 9.1: VDRL tiles

Use: To screening test for detecting syphilis.

To diagnose neurosyphilis by detecting the CSF antibodies.

VDRL tile is rotated on **VDRL rotator** (Fig. 9.2) at **180 rpm for 4 min**.

Positive VDRL test: Titres of ≥1:8 dils are considered significant.

FAQs

1. *Other non-treponemal tests are*:
   - RPR: Rapid plasma reagin test—to monitor the response to treatment in syphilis.
   - USR: Unheated serum reagin test.
   - TRUST: Toluidine red unheated serum test.
2. *Gold standard test to diagnose syphilis*: TPHA (*Treponema pallidum* hemagglutination test.
3. VDRL test is positive in 78% cases of primary syphilis, 100% in secondary syphilis and 71% in late syphilis.

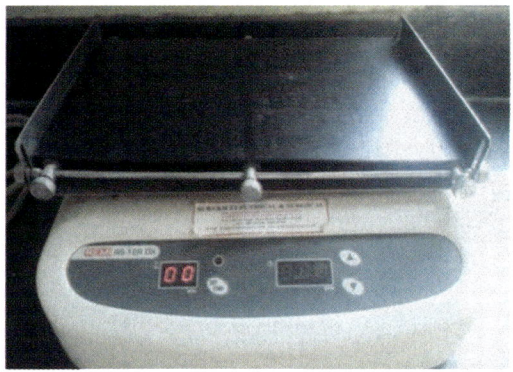

Fig. 9.2: VDRL rotator

Widal Rack (Figs 9.3 to 9.6)

Use: To perform the widal test for diagnosis of enteric fever caused by *Salmonella typhi*. It comes positive mainly in 2nd wk of infection.

Type of test: Tube agglutination test.

Widal test is named after Fernand Widal who developed this test.

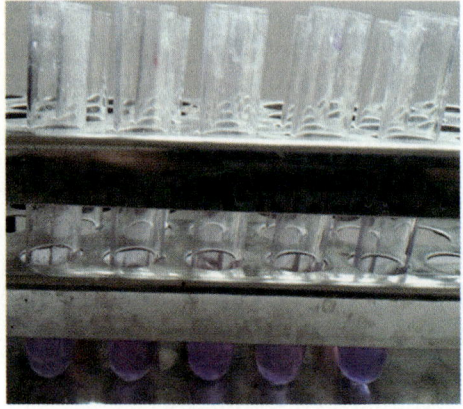

Fig. 9.3: Widal rack

Fig. 9.4: TO +ve chalky granules

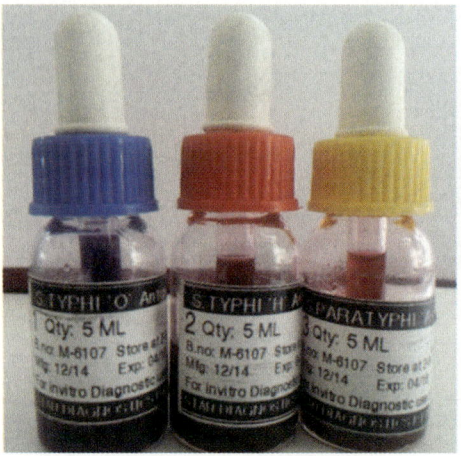

Fig. 9.5: Widal antigens

Fig. 9.6: TH +ve cotton wooly flakes

Antigen used

- Antigen (somatic antigen) of *S. typhi* (TO)
- H antigen (flagellar antigen) of *S. typhi* (TH)
- H antigen (flagellar antigen) of *S. paratyphi* A(AH)
- H antigen (flagellar antigen) of *S. paratyphi* B(BH)

FAQs

1. *Tubes used for O antigen*: Round bottom tubes also called as Dreyer's tubes.
2. *Tubes used for H antigen*: Conical bottom tubes also called as Felix tubes.
3. *Positive titres for O agglutinins*: >1:100 is considered significant. O agglutination will appear as compact chalky clumps (disc-like pattern).
4. *Positive titres for H agglutinins*: >1:200 is considered significant. H agglutination will appear as large loose fluffy cotton wooly flakes.
5. *Negative test*: Appear as button formation, if agglutination does not take place due to deposition of the antigen.

Rapid Immunochromatographic Test (Fig. 9.7)

Use: For diagnosis of:

1. HBsAg antigen detection
2. Malaria antigen detection
3. Dengue NS1 Ag detection and also for the detection of IgG and IgM antibodies
4. HIV antibodies detection
5. Chickungunya IgM antibodies detection
6. Leptospira IgG and IgM antibodies detection

Positive test: Shows the presence of capture line and control line.

Negative test: Shows the presence of only control line.

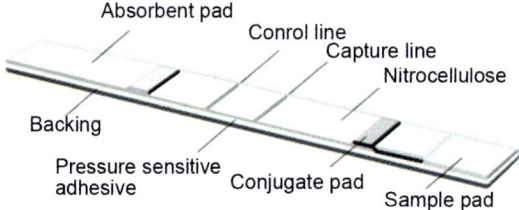

Fig. 9.7: Rapid immunochromatographic test

FAQs

1. *Hepatitis B virus (ds DNA virus)—HBsAg detection* (Fig. 9.8):

- **HBsAg antigen** is also called as Australian antigen, forms the lipoprotein envelope, appears in the serum 1–10 wks after an acute exposure to HBV and is the first marker to appear.
- HBeAg—tells infectivity or replication of the virus.
- Detection of anti-HDc antibodies—sole marker for hepatitis B infection during the window period.
- Hepatocellular carcinoma due to HBV is the vaccine preventable cancer.

2. *Malaria antigen detection* (Fig. 9.9): In general, the **blood specimen** (2 to 50 µL) is either a finger-prick blood specimen, **anticoagulated blood**, or plasma, and it is mixed with a

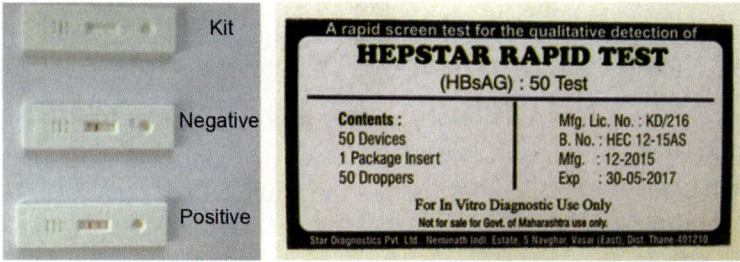

Fig. 9.8: Rapid test for Hepatitis B

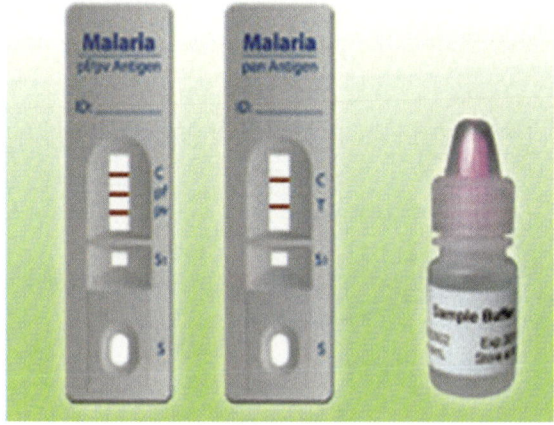

Fig. 9.9: Rapid malarial test

buffer solution that contains a hemolyzing compound and a specific antibody that is labeled with a visually detectable marker such as colloidal gold. If the target antigen is present in the blood, a labeled antigen/antibody complex is formed and it migrates up the test strip to be captured by the pre-deposited capture antibodies specific against the antigens and against the labeled antibody (as a procedural control). A washing buffer is then added to remove the hemoglobin and permit visualization of any colored lines formed by the immobilized antigen–antibody complexes (Fig. 9.10).

3. *Dengue NS1 Ag detection and also for the detection of IgG and IgM antibodies* (Fig. 9.11):

- *NS1*: A multimer forming glycoprotein secreted from infected cells and has role in replication. Thus NS1 antigens can be detected up to 9 days from the onset of infection.

- These kits generally have higher sensitivities for IgG detection but lower sensitivities for IgM detection and also show various specificities.

- *In primary infection*: IgM antibody increases by 5th day after onset of fever and IgG antibody increases by 7th day.

- *In secondary infection*: High levels of IgG are detectable even in the acute phase and they rise in next 2 wks of onset of fever.

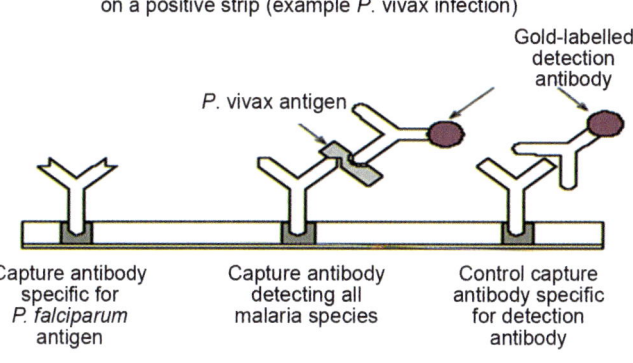

Schematic representation of immunologic reaction on a positive strip (example *P. vivax* infection)

Gold-labelled detection antibody

P. vivax antigen

Capture antibody specific for *P. falciparum* antigen

Capture antibody detecting all malaria species

Control capture antibody specific for detection antibody

Fig. 9.10: Mechanism of rapid test of malaria

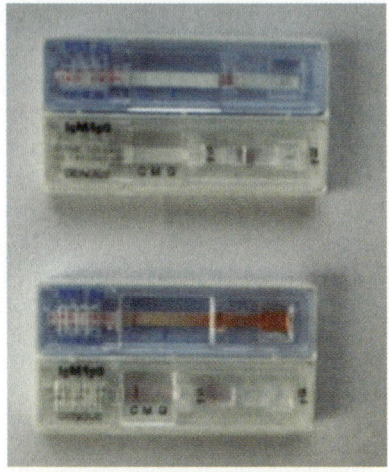

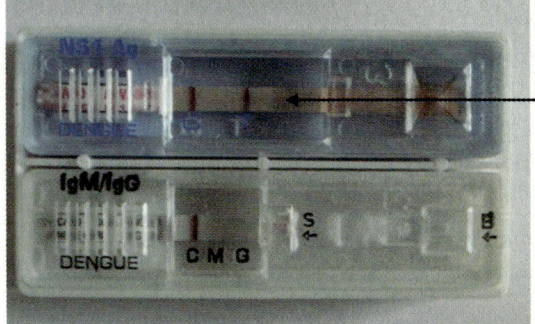

NS1 Ag positive

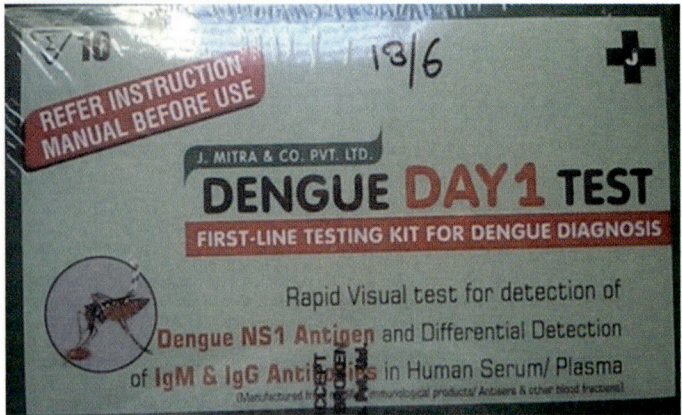

Fig. 9.11: Dengue rapid test

- *Vectors*: Mosquitoes of *Aedes* genus include—*Aedes aegypti, Aedes albopictus.*
- *Recent vaccine*: The world's first dengue vaccine against all 4 serotypes—Dengvaxia was approved by Mexico, which can be used against all 4 dengue virus serotypes for patients aged 9–45 yrs in areas where dengue is endemic.

4. *HIV antibodies detection* (Fig. 9.12):
- HIV antibodies appear in the serum 2–8 weeks after infection but usually are detectable after 3–12 weeks.
- *Screening tests* for HIV include: ELISA (used in blood screening in blood banks as a primary test) and rapid diagnostic tests.
- *Supplemental tests* include: Western blot, line immunoassays, immunofluorescent and radioimmunoassays.
- *Confirmatory test*: Western blot.
- *Detection of p24 antigen and PCR*: Used during window period (when no antibodies are present in the serum) to diagnose HIV infection.

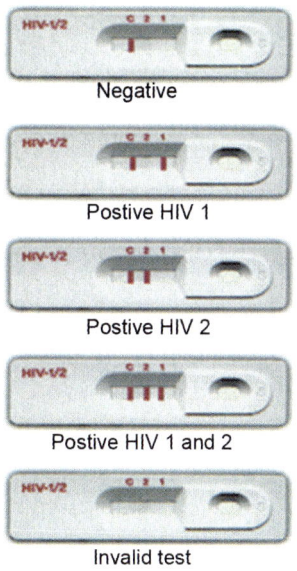

Negative

Postive HIV 1

Postive HIV 2

Postive HIV 1 and 2

Invalid test

Fig. 9.12: HIV rapid test

- *HIV tri-dot test – also called as dot blot assays* (Fig. 9.13): These tests use the recombinant or synthetic peptides spotted on nitrocellulose paper with an absorbant pad underneath. This whole assembly is enclosed in a plastic cassette.
- *Different strategies of HIV* (Fig. 9.14): Testing under NACO guidelines (National AIDS Control Organization):
- *Type II strategy is of two types IIA and IIB*:
  - IIA: For sentinel surveillance
  - IIB: For symptomatic HIV patient
- *Type III strategy* is for the **asymptomatic** HIV patient

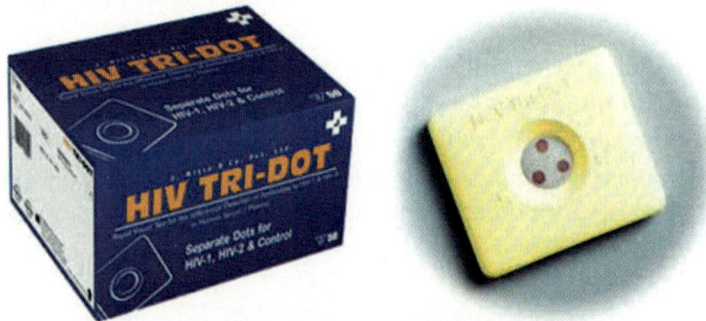

Fig. 9.13: HIV tri-dot rapid test

HIV Testing Strategies

Testing Strategy	Objective of Testing	Type of Testing	Place of Testing
I	Transfusion/ Donation safety	Mandatory	Blood Bank
II	Surveillance	Unlinked Anonymous	Designated laboratories
III	Diagnosis of patients at ICTC	Voluntary • Counselling • Informed Consent • Confidential	ICTC

Testing Related to HIV   23   NACO

Fig. 9.14: HIV testing strategy

- *Blood bank*: ELISA is preferable due to high sample load
  Note: To declare a patient HIV positive by:
    - Strategy I—one test should be positive
    - Strategy II—two tests should be positive
    - Strategy III—three tests should be positive.

5. *Chickungunya IgM antibodies detection* (Fig. 9.15):
- *Family*: Togaviridae and genus—Alphavirus
- *Transmitted by*: Aedes aegypti and *Aedes albopictus*.
- *Clinical features include*: Fever, rash, severe arthralgias and crippling arthritis and hemorrhagic manifestations. The severe arthralgia lead to the **"doubling up"** of the patient.

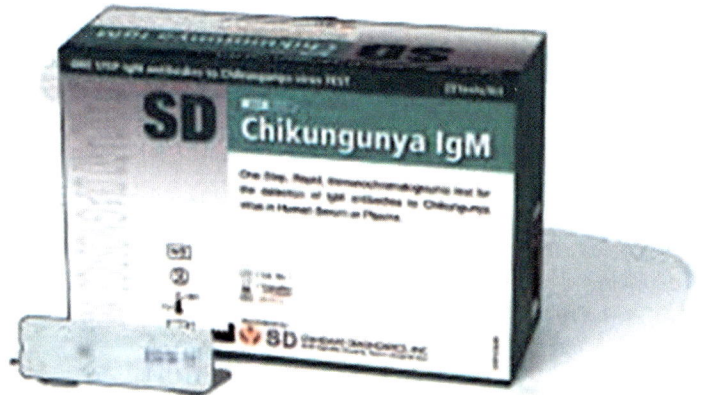

Fig. 9.15: Chikungunya rapid test

6. *Leptospira IgG and IgM antibodies detection* (Fig. 9.16):
- *Confirmatory test/gold standard*: Serovar specific test called as **MAT (microscopic agglutination test)**.
- *Leptospira has 2 species*:
  i. *L. interrogans*—pathogenic to humans. It compises of 25 serogroups with 250 serotypes.
  ii. *L. biflexa*—has 38 serogroups with 65 serovars.
- *Modes of transmission*: Contact with water, moist soil contaminated with animal urine.
- Associated with various epidemiological determinants like exposure to rodents urine, rainfall and rice fields.

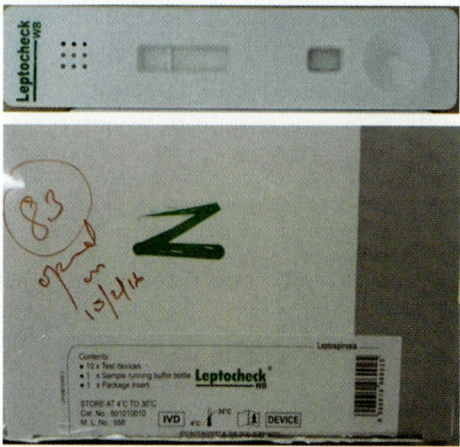

Fig. 9.16: Leptospira rapid test

Latex Agglutination Test

Principle: By attaching soluble antigens to the surface of carrier particles (polystyrene latex particles—carbon particles of size of 0.81 mm), it is possible to convert precipitation tests into agglutination tests, which are more convenient and more sensitive for the detection of antibodies.

Examples of latex agglutination tests detection of ASO, CRP, RA factor, HCG, etc.

Latex agglutination tests/**passive agglutination tests** where the latex particles are coated with the antigen, for the detection of antibodies present in the patient serum. These include: ASO titer detection.

Latex agglutination tests/**reverse-passive agglutination tests** where the latex particles are coated with the antibodies. These include:

1. CRP titer detection
2. RA factor detection.

Antistreptolysin O Titer (ASO Titre) Detection (Fig. 9.17)

Serological tests done to diagnose *streptococcal* infection: ASO titers for diagnosis of acute rheumatic fever, ASO titers >200 IU/ml considered positive when done by latex agglutination test.

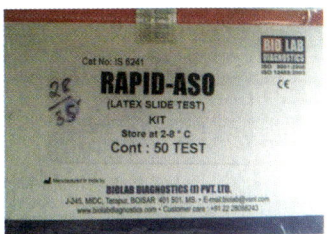 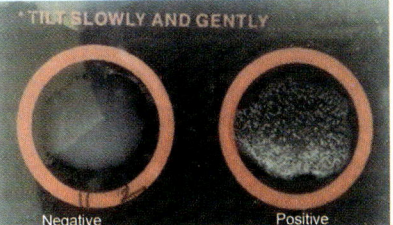

Fig. 9.17: ASO test

CRP Titer Detection (Fig. 9.18)

CRP (C-reactive protein) is an abnormal protein that appears in the blood in the acute stages of various inflammatory disorders but is undetectable in the blood of healthy persons.

The name of the protein was derived from the fact that it forms a precipitate with the non-type specific somatic C-polysaccharide of the *Pneumococcus*.

It is also increased in sepsis by any bacteria and acute tissue injury (acute myocardial infarction).

It is most commonly used as a prognostic marker rather than diagnostic test.

CRP levels: **0.6 mg/dl considered as positive** when done by latex agglutination test.

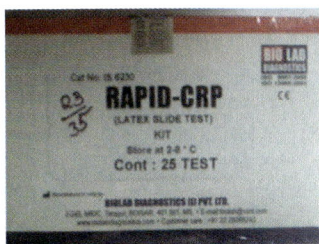 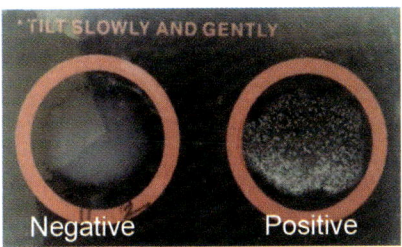

Fig. 9.18: CRP test

Rheumatoid Arthritis (RA Factor) Detection

Serological tests done to diagnose rheumatoid arthritis (Fig. 9.19).

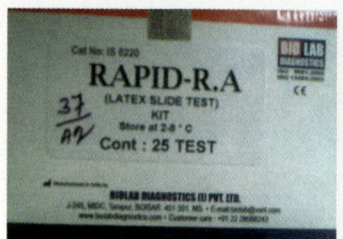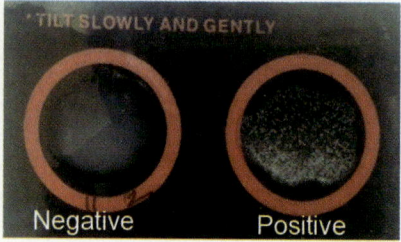

Fig. 9.19: RA test

Principle: It is a reverse passive agglutination reaction where the latex particles are coated with the anti-IgM antibodies, for the detection of IgM antibodies present in the patient serum.

RA levels for diagnosis of rheumatoid arthritis, **≥8 IU/ml** considered **positive** when done by latex agglutination test.

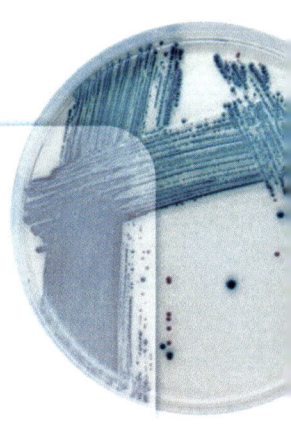

Chapter 10

Vaccines—Spots

NATIONAL IMMUNIZATION SCHEDULE

Expanded program of immunization (EPI) and universal immunization program (UIP). Following the WHO recommendation, India introduced six vaccines under the expanded program of immunization (EPI) in 1978 to reduce child mortality. These were Bacillus Calmette-Guerin (BCG), TT, DPT, DT, polio, and typhoid vaccines.

Subsequently, in 1985, the Indian Government included Measles vaccination and launched the universal immunization program (UIP) and a mission to achieve immunization coverage of all infants and pregnant women by the1990s.

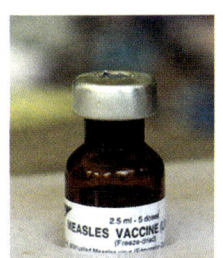
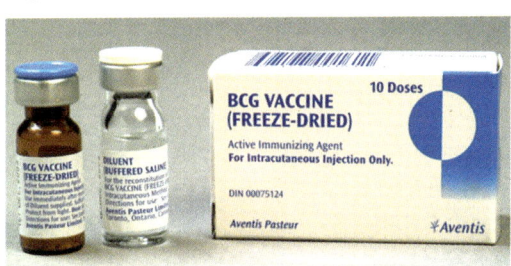
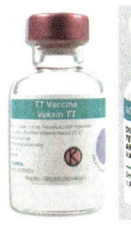
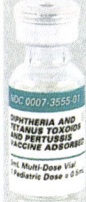
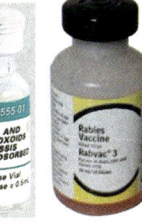
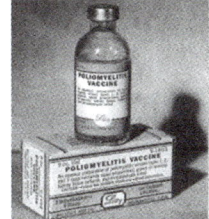

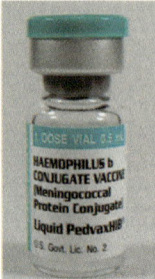
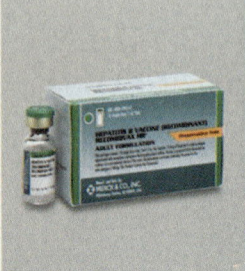
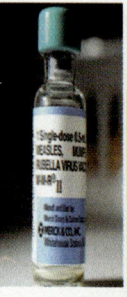
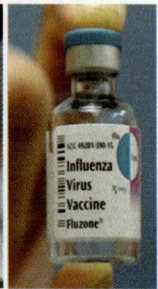

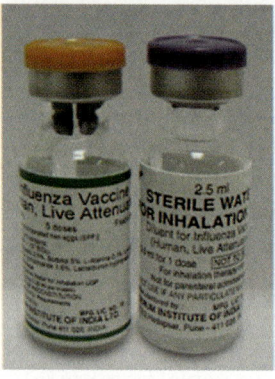
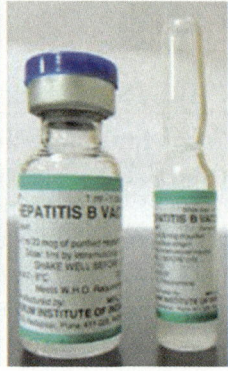
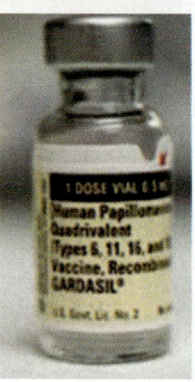

National immunization schedule

Age	Vaccines
Birth	BCG, OPV0 (for institutional deliveries)
6 weeks	DTwP1, OPV1, HepB1, Hib1
10 weeks	DTwP2, OPV2, HepB2, Hib2
14 weeks	DTwP3, OPV3, HepB3, Hib3
9–12 months	Measles
16–24 months	DTwP B1, OPV4, MMR$
5–6 years	DTwP
10 years	TT
16 years	TT
Pregnant women TT1 (early in pregnancy)	
TT2	(1 month later)
TT booster	(if vaccinated in past 3 years)
Vitamin A	9, 18, 24, 30 and 36 months

TYPE, DOSE AND SIDE EFFECTS OF VACCINE

Vaccine	Full name	Type	Dose	Side effect
BCG	Bacillus Calmette Guerin Vaccine	Live attenuated	0.1 ml	• Regional adenitis • Disseminated BCG infection • Ulcer
OPV	Oral polio vaccine Vaccine	Live attenuated	ORAL	• Vaccine induced paralysis
DTwP	Diptheria Tetanus whole cell Pertussis	Diphtheria inactivated	0.5 ml I.M	
		Pertussis inactivated Tetanus Toxoid		• **Encephalopathy**
Hep B	Hepatitis B vaccine	Recombinant	0.5 ml <19 yrs 1 ml>20 yrs IM	• Anaphylaxis
Measles	Measles	Live attenuated vaccine	0.5 ML Subcutaneous	• Purpura
TT	Tetanus toxoid	Toxoid	0.5 ml	• Pain • Induration
MMR	Measles Mumps Rubella	Live attenuated	0.5 ml SC	• Fever • Malaise
Hib	Haemophilus influenza type B	Conjugate	0.5 ml IM	• Fever • Malaise
Rabies		Human diploid cell culture vaccine	1 ml	• Fever • Pain • Malaise

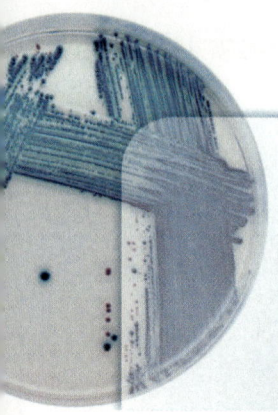

Chapter 11

Antimicrobial Susceptibility Testing (AST)

Clinical Laboratory Standard Institute (CLSI), United States of America, and the European Committee on Antimicrobial Susceptibility Testing (EUCAST) have published approved protocol of antimicrobial susceptibility testing either by disk diffusion assay or broth microdilution.

1. Disk Diffusion Method

It is done by Kirby-Bauer method recommended for anti-microbial susceptibility testing by disk diffusion by CLSI (Figs 11.1 to 11.3).

Types of medium used

1. *For bacteria*: Mueller-Hinton agar (MHA); pH (7.2–7.4), 4 mm depth, uniformly spread out in 90 mm petridish.

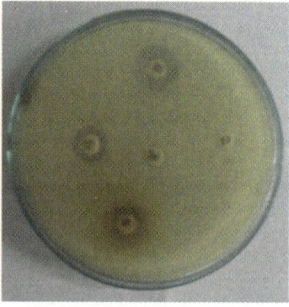

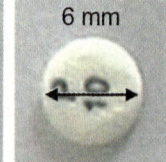

6 mm

Fig. 11.1: Uninoculated MHA   **Fig. 11.2:** Antibiotic disk   **Fig. 11.3:** Plain MHA

2. *For yeast*: Glucose methylene blue; pH (7.2–7.4), 4 mm depth, uniformly spread out in 90 mm petridish.

Size of inoculums matched with: 0.5 McFarland

Incubation time and temperature for AST plate: 35°–37°C for 16–18 hrs.

Antimicrobial agent (antibiotic disks)

1. Concentration and potency of antibiotic disks checked with standard strains
   - *Staphylococcus aureus* ATCC 25923
   - *Escherichia coli* ATCC 25922
   - *Pseudomonas aeruginosa,* ATCC 27853
2. *Disc size*: 6 mm
3. *Storage*: Short term: 2–8°C; Long term: –20°C

Procedure of AST: Sterile cotton swab stick was dipped into the broth of each isolate and rotated firmly against upper inner side wall of the tube to express excess fluid. Lawn culture was made by streaking the swab evenly in 3 planes on to the surface of the 90 mm glass petridish containing Mueller-Hinton agar (4–5 mm in depth, dried for 30 min).

Plates are then allowed to dry for 3–5 min at room temperature not more than 15 min.

Application of antibiotic disks: Antibiotic disks were placed on the lawn culture of test isolate with the help of sterile forceps and gently pressed to ensure contact with the medium. The disks were placed evenly so that:
- They are no closer than 24 mm center-to-center
- No closer than 18 mm from edge of the petridish
- Not more than 5 disks were applied on a plate.

Advantage of Kirby-Bauer method: The agar-based method is a reliable alternative to microdilution method.

Disadvantage of Kirby-Bauer method: The detection of resistance by agar-based methods correlates poorly with the detection of resistance by the microdilution method.

2. Macrodilution Method (Fig. 11.4)

Macrobroth dilution method requires tubes for testing the anti-microbial susceptibility of the organism. This procedure is **based on two-fold serial dilutions** of the antibiotic.

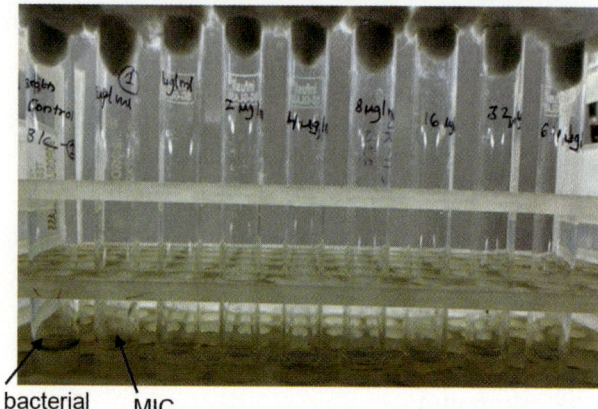

Normal bacterial   MIC
broth/suspension

Fig. 11.4: Macrodilution test

Medium defined for bacteria: Mueller-Hinton broth

For fungus: RPMI-1640 with MOPS

Optimum incubation was determined to be 35°C for 48 or 72 hrs depending on species.

Use: To determine the minimum inhibitory concentration (MIC). It is defined as the lowest dilution that resulted in zero visible growth for antibiotic or the lowest concentration of the antimicrobial agent that inhibits the growth.

3. Microdilution Method (Fig. 11.5)

Done in 96-well microtiter plate, as is also used for the ELISA testing. This gives the minimum inhibitory concentration values (**MIC values**), which are defined as the lowest dilution that resulted in **zero visible growth**.

FAQs

Minimum bactericidal concentration (MBC): The lowest concentration of the antimicrobial agent that is required to **kill** a particular bacterium. It is determined from the broth dilution MIC test by further subculturing to the agar plates that do not contain these agents. Thus MBC is defined by **determining the lowest concentration of antimicrobial agent that reduces the viability of the initial bacterial inoculums by ≥99.9%.**

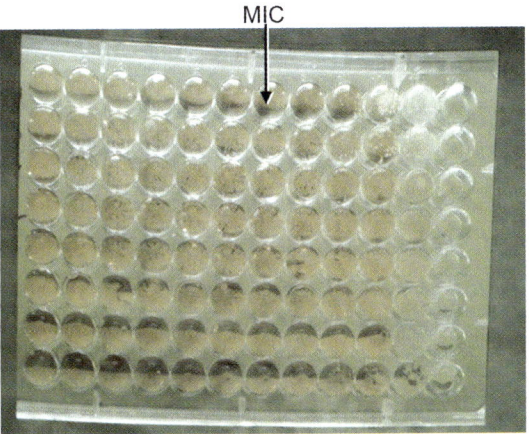

Fig. 11.5: Microdilution test

4. E Test (Fig. 11.6)

E test is based on a combination of the concepts of dilution and diffusion tests. Like dilution methods, E test directly quantifies antimicrobial susceptibility in terms of discrete MIC values.

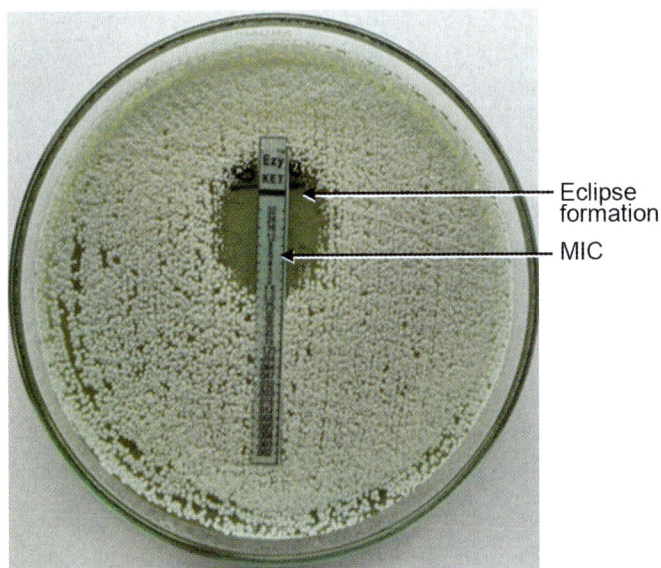

Fig. 11.6: E test

As E test consists of a predefined and continuous concentration gradient, the MIC values obtained can be more precise than values from conventional procedures based on discontinuous two-fold serial dilutions.

E test consists of a thin, inert and non-porous plastic strip (5 × 60 mm). One side of the strip is calibrated with MIC values in µg/ml and a two-letter code designates the identity of the drug. A predefined and exponential gradient of the dried and stabilized antimicrobial agent is immobilized on the other surface of the strip. The continuous gradient covers a concentration range corresponding to 15 two-fold dilutions in a conventional dilution procedure.

When applied on to an inoculated agar plate, there is an immediate release of the agent from the plastic surface into the agar matrix. A predefined, continuous and stable gradient of the drug concentrations is created directly underneath the strip.

Interpretation: After incubation, whereby growth becomes visible, a symmetrical inhibition **ellipse** centered along the strip is seen. The zone edge intersects the strip at the MIC value.

Factors affecting antifungal susceptibility testing
1. *End-point determination*: This is an important source of variability among laboratories in the testing of agents that are generally bacteriostatic/fungistatic and usually do not have distinct end-points on MIC testing and thereby introduce subjective interpretations of susceptibility results contrast from bacteriocidal/fungicidal drug and there is not much problem in interpreting end-points.
2. *Inoculum size*: Increasing the size of inoculum can drastically increase MIC for most drugs tested.
3. *Incubation*: MIC tends to increase with longer incubation periods.
4. *Media*: Also the pH of the test affects the MIC interpretation, like lower pH of the test medium is associated with higher MIC of most antifungal agents.

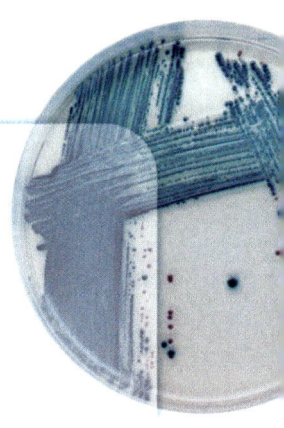

Chapter 12

Virology—Spots

Egg (Fig. 12.1)

Most commonly used egg: White leghorn egg.

Other eggs used: Duck egg for rabies virus, hens egg.

Egg: Hens egg is used for viral isolation by inoculating in various layers of the egg (Fig. 12.2).

Fig. 12.1: Egg

Layers of egg	Viruses isolated	Bacteria isolated
Chorioallantoic membrane	Vaccinia Variola HSV-2 Fowlpox Cowpox	Brucella spp Borrelia spp Leptospira spp Listeria spp
Amniotic cavity	Influenza virus A, B and C Mumps virus	—
Allantoic cavity	Influenza virus A, B-primary isolation and also for antigen preparation	Listeria spp (vaccine preparation)
Yolk sac	Influenza virus HSV	Chlamydia spp Rickettsiae spp

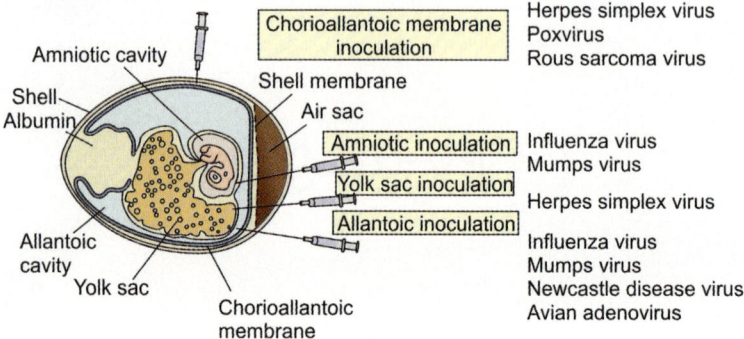

Fig. 12.2: Egg inoculation—different sites

Tissue Culture Bottles (Fig. 12.3)

Use

1. To isolate the viruses.

2. To see the various cytopathic effects (CPE) produced by the virus on various cell lines with the help of inverted microscope.

Various cell lines used

1. *Primary cell lines*:

   • African green monkey cell lines

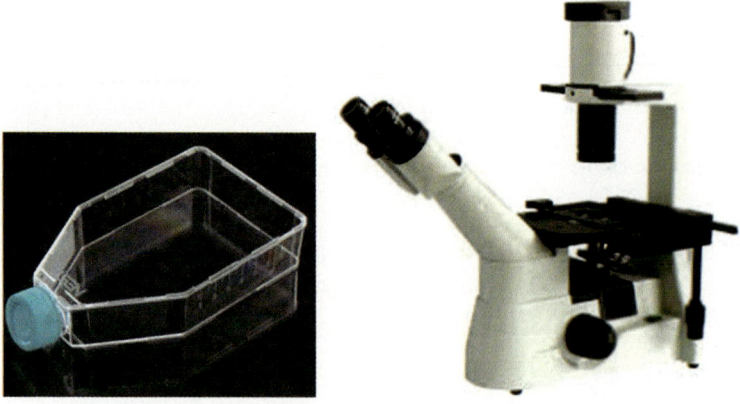

Fig. 12.3: Tissue culture bottle and inverted microscope

- Rhesus monkey kidney cell lines
- Cord blood mononuclear cells lines
2. *Diploid cell lines*:
   - WI-38 cell lines
   - MRC-5 cell lines
3. *Continuous cell lines*:
   - HeLa cell lines
   - Hep-2 cell lines
   - MDCK cell lines

Viral Transport Medium (VTM) (Fig. 12.4)

Use: Transport of specimens suspected of having viruses for their identification.

Composition
1. *Proteins*: To stabilize the viruses
2. *Antibiotics*: To prevent bacterial and fungal growth
3. *Buffer*: To maintain pH—6.8–7.6
   Temperature at which specimen is transported in VTM-4°C.

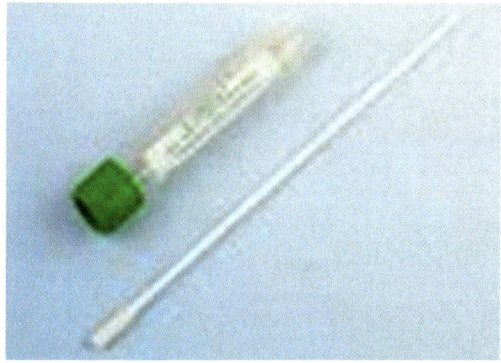

Fig. 12.4: Viral transport media

Minimal Essential Medium (Fig. 12.5)

Use: To **maintain** the virus culture in cell lines.

Composition
1. Earle's or Hanks balanced salt solution
2. Essential amino acids

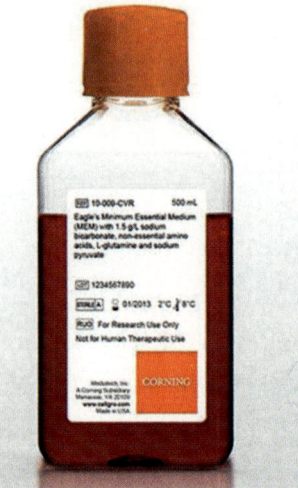

Fig. 12.5: Eagle minimum essential media

3. Vitamins
4. Buffer
5. Antibiotics
6. Serum—**fetal bovine serum (2%)**

FAQs

DNA viruses	
Double-stranded DNA viruses **(dsDNA viruses)**	Poxvirus
	Herpes virus
	Hepatitis B virus
	Papilloma virus
Single-stranded DNA viruses **(ssDNA viruses)**	Parvovirus B-19 virus

RNA viruses	
Double-stranded RNA viruses **(dsRNA viruses)**	Reovirus
	Rotavirus
Single-stranded RNA viruses **(ssRNA viruses)**	Filoviruses
	Influenza virus
	Parainfluenza virus
	Picorna virus

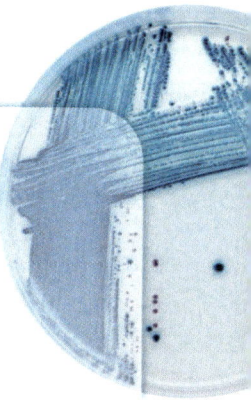

Chapter 13

Stool Examination

Common preservative used in stool sample (for parasitological examination)
 i. 5–10% formalin
 ii. Merthiolate iodine formalin solution (MIF)
 iii. Sodium acetate acetic acid-formalin (SAF) fixative
 iv. Polyvinyl alcohol (PVA) fixative

Examination of stool includes:
1. Macroscopic or gross examination
2. Microscopic examination

MACROSCOPIC OR GROSS EXAMINATION

Quantity: In intestinal amebiasis, the stools tend to be voluminous whereas in bacillary dysentery due to *Shigella,* the stools are scanty in quantity.

Consistency and form
• Normal stools are well formed.
• In diarrhea and dysentery, the stools are semisolid or watery in nature.
• In cases of malabsorption of fats, the stools are pale bulky and semisolid.

Color
• Normal stools are light to dark brown in color due to the presence of stercobilinogen which is a product of bilirubin metabolism.

- In cases with bleeding into the intestinal tract, the stools become dark tarry in nature due to the formation of acid hematin, if the bleeding is in the small intestines.
- In the bleeding in large intestines or rectum, the blood may be bright red.
- In cholera, the stools have a rice water appearance as there is no fecal matter and there is presence of flakes of epithelial cells in it.
- In biliary tract obstruction, the stools may be clay-colored due to absence of stercobilinogen.

Odor
- The fecal odor of stools may become offensive in conditions like intestinal amebiasis.
- In cases of bacillary dysentery and cholera, the stools are not foul smelling due to the absence of fecal matter.

Blood: Blood should be noted in stools, if present as it is indicative of ulceration or presence of any other pathology like malignancy.

Mucus: Mucus is present in certain conditions like amebic or bacillary dysentery.

Parasite: Stools may contain adult helminths. Nematodes, like *Ascaris*, are easily visible as their size is large. Hookworms and proglottids of cestodes may also be present. These may be visible to the naked eye.

MICROSCOPIC EXAMINATION

The laboratory diagnosis of most parasitic infections is by the demonstration of ova of the parasite in the stools of the infected person. The stool is collected in a clean wide-mouthed container. The stool can be examined by the following techniques.

a. Saline wet mount examination
b. Iodine preparation
c. Buffered methylene blue stain—nuclear stain in *E. histolytica*
d. Cellophane tape test—NIH swab
e. Concentration techniques

a. *Saline wet mount examination*: The stool is emulsified in normal saline and a large drop is placed on a glass slide and is then covered with a cover slip. This is then examined under a light microscope using low and high power objective lens. The thickness of the film should be such that one is able to see the printed letters of the newspaper through it.

b. *Iodine preparation*: Iodine preparation leads to better visualization of morphological details of ova and cysts as it stains the glycogen in them.
- It, however, has the disadvantage that the live trophozoites of *Entamoeba histolytica* cannot be seen as the iodine kills it.
- Lugol iodine is used for stool examination

Preparation of Lugol iodine
- Two grams of potassium iodide are mixed in 100 ml distilled water and then one gram iodine crystals are added and it is shaken vigorously.
- The solution is then filtered into a dark glass bottle and kept away from light.

c. *Buffered methylene blue stain*: It is used for the nuclear stain in *E. histolytica*.

d. *Cellophane tape test*: NIH swab (National Institute of Health) is used to pick up the ova of *Enterobius vermicularis* from the perianal area.

e. *Concentration methods*: Two types of concentration techniques are used for stool examination:
  i. Sedimentation technique
  ii. Floatation technique

  i. *Sedimentation technique*:
     1. Formol ether technique
     2. Simple gravity sedimentation technique
  ii. *Floatation technique*:
     1. Zinc sulfate
     2. Saturated salt solution

Saturated salt solution floatation technique: A small amount of stools are emulsified in saturated salt solution in a wide mouth container. When the stool becomes homogenous a few drops of saturated saline are mixed and more saturated saline solution is added.

A glass slide is placed across the receptacle in such a way that the slide is in contact with the surface of the saturated solution. If the fluid is not in touch with the slide then more saturated saline should be added till the fluid level touches the slide.

The slide is then left in place for 15 minutes. After this, the slide is gently lifted off the container and turned upside down carefully ensuring no fluid from the slide is spilled.

A cover slip is placed over the fluid on the slide and it is examined under a microscope.

Principle of floatation technique: The specific gravity of ova and cysts is less and thus will float to the top of the saturated salt solution where it will stick to the under surface of the glass slide.

Example of egg that float in saturated salt solution
- *Fertilized egg of Ascaris lumbricoides*
- *Ancylostoma duodenale*
- *Trichuris trichiuria*, Mnemonic: "**FATEH**"
- *Enterobius vermicularis*
- *Hymenolepsis nana*

Principle of formol ether concentration technique (Fig. 13.1)

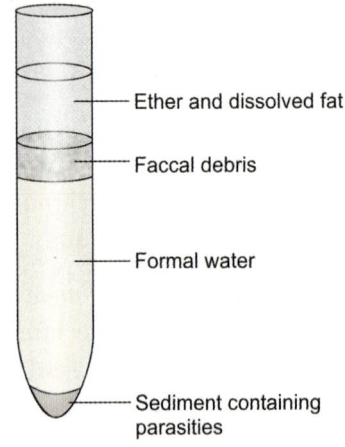

  i. Formalin fixes the egg cyst and larva and ether removes fatty material from stool so egg cyst and larva are better visualized.

Ether and dissolved fat

Faccal debris

Formal water

 ii. Four layers are formed after centrifugation of this procedure.

iii. Saline mount and wet mount are made from sediment at the bottom of centrifuge tube.

Sediment containing parasities

Fig. 13.1: Four layers of formol ether concentration method

COMMON PARASITES FOUND IN STOOL SAMPLE

Egg of *Trichuris trichiura* (Fig. 13.2)

Bile-stained, barrel-shaped with mucous plug at each pole, measuring 50–54 µm × 22–23 µm, **float** in saturated common salt solution.

Single host: Man

Habitat: Large intestine

Infective stage: Embryonated eggs containing a **rhabditiform larva** (250 µm).

Disease caused: Whipworm dysentery leading to rectal bleeding.

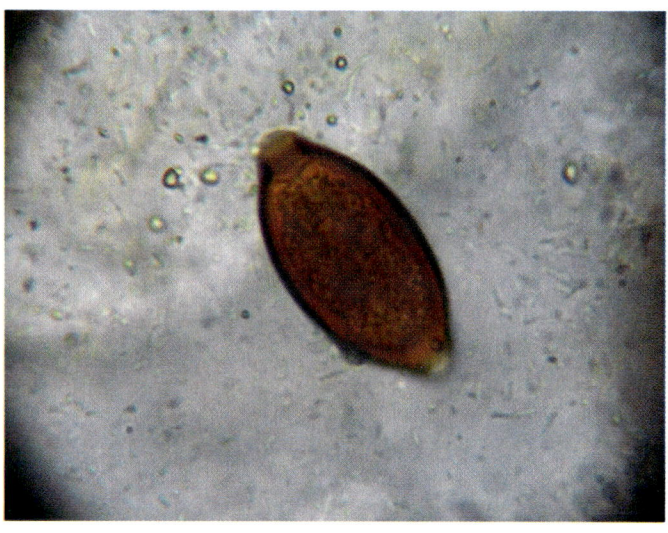

Fig. 13.2: Egg of *Trichuris trichiura*

Egg of Hookworm (Fig. 13.3)

Non-bile-stained, oval or elliptical, surrounded by a transparent shell, measuring 60 µm × 40 µm, contains a segmented ovum with usually 4 blastomeres and a clear space between the egg shell and segmented ovum, float in saturated common salt solution.

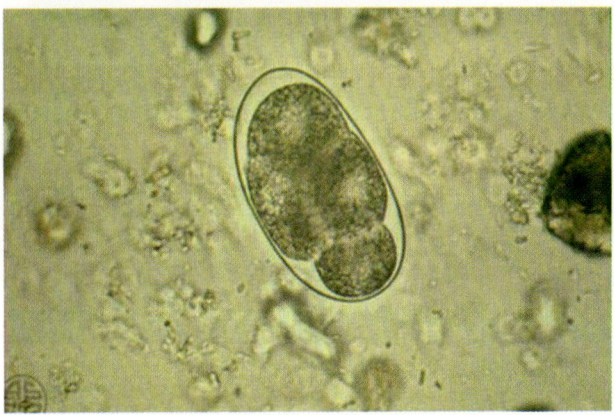

Fig. 13.3: Egg of hookworm

Single host: Man

Habitat: Small intestine

Infective stage: **Filariform larva** (500–700 µm) penetrating the skin.

Disease caused: Dermatitis or ground itch, pulmonary lesions like bronchitis and bronchopneumonia and iron deficiency anemia.

FAQs
Worm load calculated by
  i. *Chandler index*—based on the average number of eggs per gram of feces for the entire community, it is as follows:

Egg number	Significance
<200	Hookworm not of much significance
200–250	Potential danger
250–300	Minor public health problem
>300	Important public health problem

Use of Chandler index
  • Epidemiological studies of hookworm.
  • Worm loads in different population can be compared and also the degree of reduction of egg output after mass treatment.

ii. Adult female worm produces about 2500–5000 eggs/day, thus the fecal egg count may reflect the number of adult hookworms.

Egg of *Enterobius vermicularis* (Fig. 13.4)

Other name: Threadworm/pinworm/seatworm

Slide shows: Colourless, non-bile-stained, planoconvex (flattened on one side), measuring 60 µm × 30 µm and surrounded by a thin smooth transparent shell usually containing a developed larvae, float in the saturated salt solution.

Single host: Man

Habitat: Cecum and appendix.

Fertilized Egg of *Ascaris lumbricoides* (Fig. 13.5)

Bile-stained, round to oval, measuring 60–75 µm × 40–50 µm unsegmented ovum and a clear crescentric space at each pole, float in saturated common salt solution.

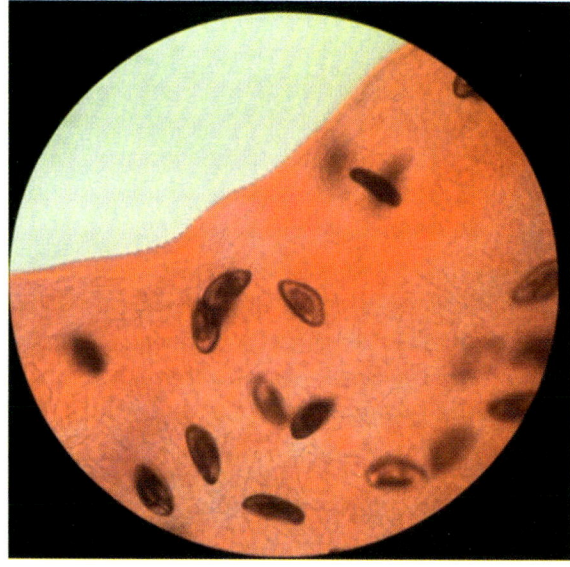

Fig. 13.4: Egg of *Enterobius vermicularis*

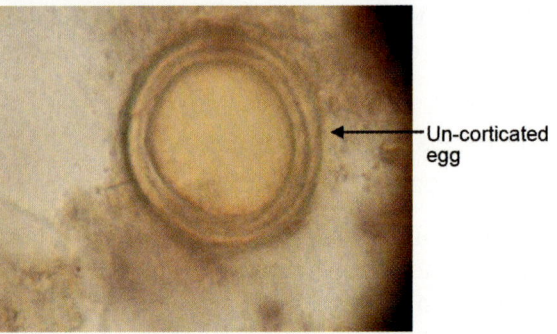

Un-corticated egg

Fig. 13.5: Fertilized egg of *Ascaris*

Consists of three layers:

1. *Outermost layer*: Coarsely mamillated albuminoid layer and if this layer is absent it is called as uncorticated egg.
2. *Middle layer*: Thick and transparent.
3. *Innermost layer*: Lipoidal vitelline membrane.

Single host: Man.

Habitat: Small intestine particularly jejunum.

Infective stage: Embryonated eggs containing a **rhabditiform larva** (250 µm).

Diseases caused: Malnutrition, intestinal obstruction, Loeffler's syndrome (migrating larva leading to allergic and hyper-sensitivity reaction in lungs).

Unfertilized Egg of *Ascaris lumbricoides* (Fig. 13.6)

Bile-stained, narrow and longer, measuring 90 µm × 55 µm, atrophied ovum and clear crescentric space at each pole is **absent**, it **sinks in saturated common salt solution.**

Consists of three layers:

1. *Outermost layer*: Thin irregular albuminoid layer.
2. *Middle layer*: Thick and transparent.
3. *Innermost layer* (lipoidal vitelline membrane): Absent.

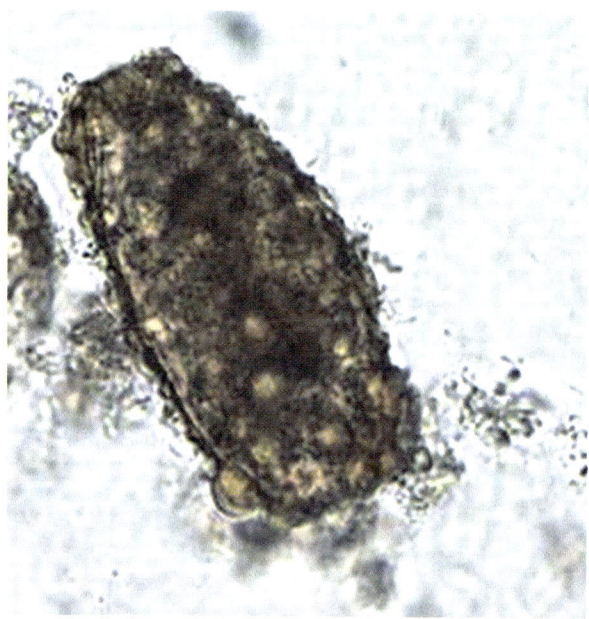

Fig. 13.6: Unfertilized egg of *Ascaris*

FAQs

1. Unfertilized egg of *Ascaris* is the heaviest of all the helminthic eggs.

2. Eggs which sink in the saturated common salt solution:
   - Unfertilized egg of *Ascaris*
   - Egg of *Taenia*
   - Egg of all flukes

3. Bile-stained eggs (mnemonic—ATT) are:
   - **A**—*Ascaris* corticated egg
   - **T**—*Taenia* egg
   - **T**—*Trichuris trichiura*

4. NOT bile-stained eggs (mnemonic—NEHA)
   - **N**—NOT bile-stained
   - **E**—*Enterobius vermicularis* egg
   - **H**ookworm eggs and *Hymenolepsis nana* egg
   - **A**ncylostoma spp.

Egg of *Taenia* (Fig. 13.7)

Bile-stained, spherical, measuring 31–43 µm, **sinks** in saturated common salt solution.

Consists of three layers:
1. *Outermost layer:* Transparent shell (represent remnants of the yolk mass).
2. *Middle layer:* Thick and **striated** embryophore.
3. *Inside embyrophore:* Hexacanth embryo (Onchosphere) with three pairs of hooklets.

Definitive host: Man

Intermediate host: Pig for *Taenia solium* and cattle in *Taenia saginata*.

Habitat: Small intestine

Infective stage: Cysticercus cellulosae (encysted larval stage in pork muscles), measuring 8–10 µm × 5 µm.

Disease caused by
1. *Adult worm:* Abdominal discomfort, anemia, intestinal disorders.
2. *Larvae:* Cysticercosis.

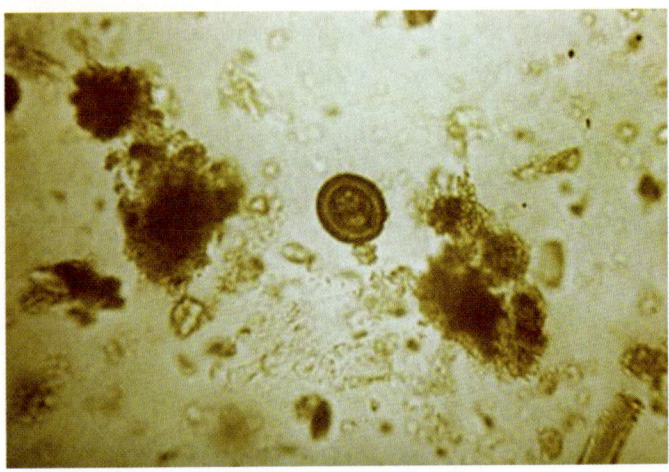

Fig. 13.7: Egg of *Taenia*

FAQs
1. Differentiation of egg of *Tinea solium* and *T. saginata*
   - Eggs of *Tinea saginata* are acid fast in nature on performing ZN staining.
2. *Drug of choice*: Praziquantal.
3. Parasitic showing **autoinfection** as mode of infection:
   - *Taenia* spp
   - *Hymenolepsis nana*
   - *Enterobius vermicularis*
   - *Strongyloides stercoralis*

Egg of *Hymenolepsis nana* (Fig. 13.8)

Non-bile-stained, spherical or oval, measuring 30–45 µm, floats in saturated common salt solution.

Consists of three layers:
1. *Outermost layer*: Transparent shell.
2. *Middle layer*: **Non-striated** embryophore and space between the embrophore and outer shell contains 4–8 polar filaments arising at either end of embryophore.
3. *Inside embyrophore*: Hexacanth embryo (Onchosphere) with three pairs of hooklets.

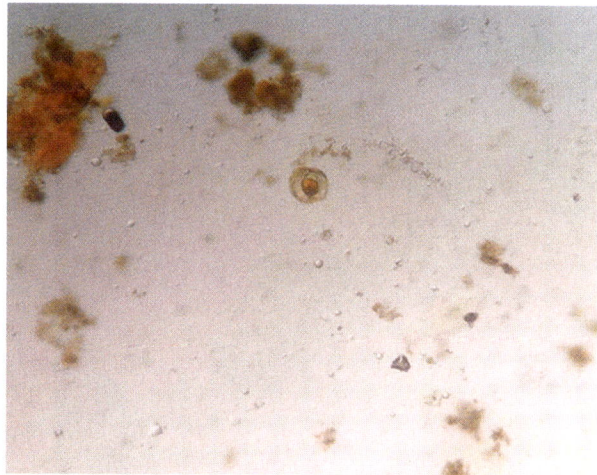

Fig. 13.8: Egg of *Hymenolepsis nana*

Single host: Man

Habitat: Small intestine

Infective stage: Embryonated egg.

FAQs
1. Only cestode to complete its life cycle in a single host.
2. *Drug of choice*: Praziquantal.

Rhabditiform Larva of *Strongyloides stercoralis* (Fig. 13.9)

Most commonly seen in stool sample.

Length—200–250 µm and width—16 µm, short mouth and **a characteristic double bulb esophagus.**

FAQs
1. It is the smallest nematode (2–3 µm × 30–50 µm).
2. Female is ovoviviparous and lays eggs by parthenogenesis which means reproduction by means of unfertilized ovum.
3. Male worm show the presence of copulatory spicules.
4. **Infective form—filariform larva** (650 µm × 10 µm) which penetrate skin and lead to serpiginous pruritic eruption, this condition is called as **larva currens (means racing larva).**

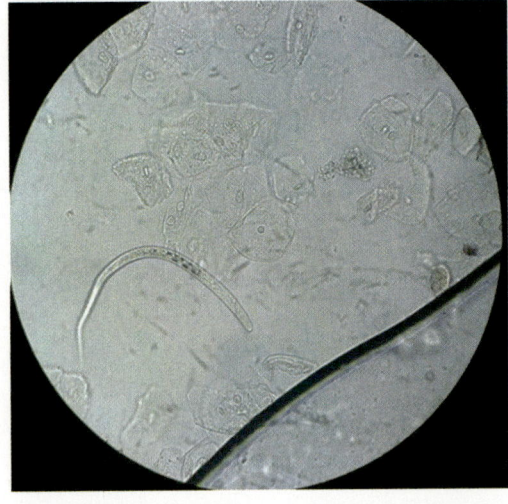

Fig. 13.9: Rhabditiform larva of *Strongyloides stercoralis*

Adult Worm of *Fasciola buski* (Fig. 13.10)

Other name: Large/giant intestinal fluke.

Slide shows the characteristic leaf-like unsegmented flat body, fleshy dark red, elongate to ovoid-shaped, the anterior end being narrow than the posterior end, measuring 20–75 μm × 8–20 μm × 0.5–3 μm with no cephalic cone.

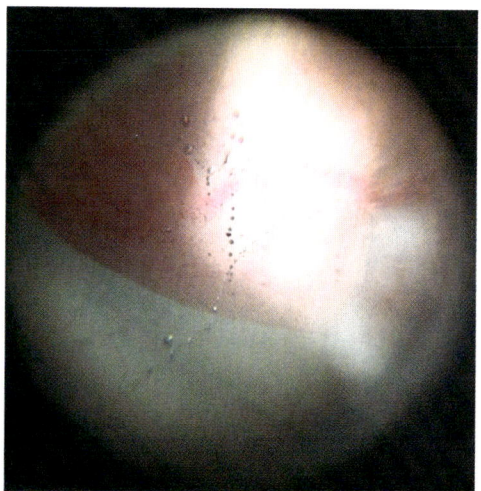

Fig. 13.10: Adult worm of *Fasciola buski*

FAQs

1. *Habitat*: Small intestine (duodenum and jejunum)
2. *Infective form*: Metacercaria encysted on acquatic vegetations.
3. *Intermediate host*: 1st is snail and 2nd is aquatic vegetation especially the seed pods of water caltrops.
4. *Eggs*: Bile-stained, **operculated** and **sinks** in saturated salt solution.

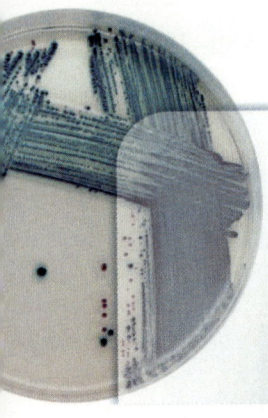

Chapter 14

Hospital Waste Management

 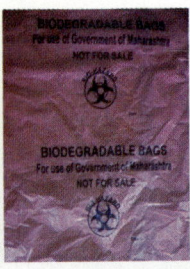  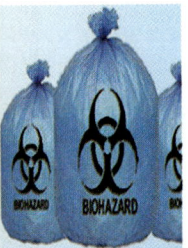

Fig. 14.1: Color coding of BWM

Bag color	Use for discarding items
Yellow	a. **Human anatomical waste:** Human tissues, organs and body parts
	b. **Animal anatomical waste:** Experimental animal carcasses, body parts, organs, tissues, including the waste generated from animals
	c. **Soiled waste: Items contaminated with blood**, body fluids like gloves, dressings, plaster casts, cotton swabs and bags containing residual or discarded blood and blood components
	d. **Expired or discarded medicines** including all items contaminated with cytotoxic drugs
	e. **Chemical waste:** Chemicals used in production of biologicals and used/discarded disinfectants
	f. **Discarded linen**, beddings **contaminated with blood** or body fluid

(Contd…)

Bag color	Use for discarding items (Contd...)
	g. *Microbiology,* biotechnology and other clinical laboratory waste *Laboratory cultures,* stocks or specimens of micro-organisms, live or attenuated vaccines, human and animal cell cultures used in research, industrial laboratories, production of biologicals, residual toxins, dishes and devices used for cultures
Red	**Contaminated waste (recyclable):** Wastes generated from **disposable items** such as tubings, bottles, intravenous tubes and sets, catheters, urine bags, syringes (without needles)
White (translucent/ puncture-proof containers)	Waste sharps including metals **Needles, syringes with fixed needles**, scalpels, blades, or any other contaminated sharp object that may cause puncture and cuts. This includes both used, discarded and contaminated sharps
Blue (Puncture-proof containers)	Glass Broken or discarded and contaminated glass
Black	Discarding non-hazardous waste

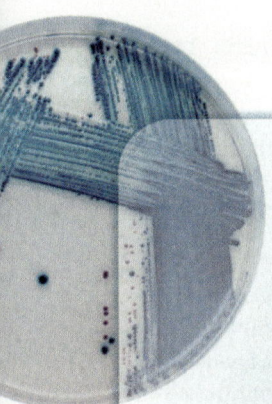

Chapter 15

Animals

Laboratory animals play a vital role in research/teaching/ diagnosis. The various animals used are as follows:

1. Rabbit
2. Guinea pig
3. Mice
4. Rat
5. Hamster
6. Monkey
7. Chimpanzee
8. Ferret
9. Sheep

Rabbit (Fig. 15.1)

- For preparation of antisera, vaccine production.
- Virulence test for *C. diphtheriae*.
- Intracerebral inoculation for rabies virus.
- Maintenance of Nichol's strain *T. pallidum*.
- Anton's test for *Listeria*.
- Ligated ileal loop for heat labile toxin ETEC (enterotoxicogenic *Escherichia coli*).
- Differentiate human and bovine strains of *M. tuberculosis*.
- Scarification of skin for propagation of *Vaccinia* virus.

Fig. 15.1: Rabbit

Guinea Pig (Fig. 15.2)

- Virulence test for *C. diphtheriae*.
- Intraperitoneal inoculation for isolation: *Yersenia pestis, Nocardia* spp, *M. tuberculosis, Clostridia* spp, *Leptospira and Streptobacillus moniliformis*.
- Sereny's test: Keratoconjunctivitis in *Shigella* and EIEC (enteroinvasive *Escherichia coli*).
- Source of complement for complement fixation test.

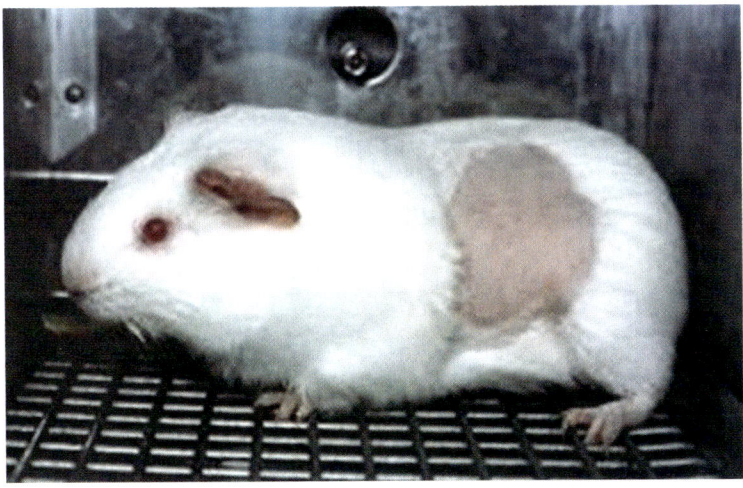

Fig. 15.2: Guinea pig

- Subcutaneous inoculation for isolation of *B. anthracis and F. tularensis.*
- To differentiating *Rrckettsial* species (endemic typhus) by Neil Mooser reaction.

Mice (Fig. 15.3)

- Intraperitoneal (IP) injection of hybridoma cells for monoclonal antibody production.
- IP route for isolation of *Streptobacillus moniliformis, Borrelia* spp, *Cryptococcus neoformans, Histoplasma capsulatum, Nocardia* spp, *T. gondi, C. immitis, B. dermatitidis.*
- IP route for culture of *pneumococci.*
- For studying pathogenesis of prions, scrapie and *lymphocytic choriomeningitis virus.*
- Intacutaneous route for propagation and isolation of *Chlamydia.*
- Isolation of *rickettsial* species.

Fig. 15.3: Mice

Suckling Mice (Fig. 15.4)

- Intragastric route for ETEC for detection of ST toxin.
- Intraperitoneal route for differentiating *Coxsackie* virus A and B.
- Intracutaneous route for *Herpes simplex virus* and Arbovirus.

Fig. 15.4: Suckling mice

Rat (Fig. 15.5)

- Intraperitoneal route for demonstration of *Borrelia, Toxoplasma gondii.*
- Subcutaneous route for *Y. pestis.*

Fig. 15.5: Rat

Hamsters (Fig. 15.6)

- Intracutaneous, intraperitoneal, or subcutaneous inoculation of *coxsackie virus* in suckling hamster.
- Intraperitoneal route for demonstration of *Leishmania donovani, Clostridia* spp, *Leptospira* spp.
- Intradermal route for demonstration of *Leishmania tropica, Leishmania braziliensis.*

Fig. 15.6: Hamster

Ferrets (Fig. 15.7)

Highly susceptible to:

- *Influenza virus* and *Canine distemper viruses*. With distemper virus infection, mortality is 100%.
- *Staphylococcal* and *streptococcal* infection.
- Toxoplasmosis.

Fig. 15.7: Ferrets

Monkeys (Fig. 15.8)

- For studying pathogenesis of Creutzfeldt-Jakob disease, Kuru, *rotavirus*, HAV.
- Intracutaneous or intraperitoneal route for demonstration of *polio* virus.
- Scarification for propagation of *vaccinia virus*.
- Intranasal instillation of *variola* virus.

Fig. 15.8: Monkey

Chimpanzee (Fig. 15.9)

- Intraperitoneal route for Hepatitis A virus.
- Intranasal inoculation for *rhinovirus.*

Fig. 15.9: Chimpanzee

Sheep (Fig. 15.10)

- Study of pathogenesis of prion disease, scrapie.
- Scarification for the propagation of *vaccinia* virus.
- Sheep blood used to make blood agar.

Fig. 15.10: Sheep

Armadillo (Fig. 15.11)

Leprosy bacilli can be grown in nine banded armadillo called as *Dasypus novemictus*.

Fig. 15.11: Armadillo

Chapter 16

Symbols of Importance

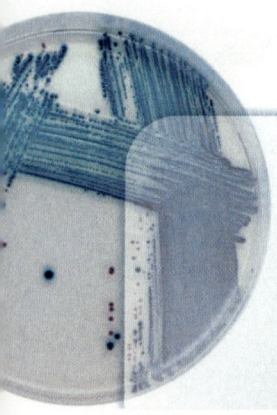

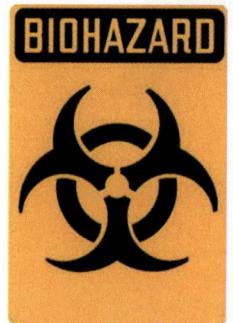

ULTRAVIOLET
RADIATION

Caution

DANGER

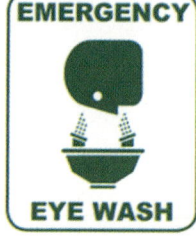

EMERGENCY

EYE WASH

EMERGENCY

SAFETY
SHOWER

DOTS

Pura Course + Pakka ilaaj

National AIDS Control Organisation

India's Voice against AIDS

National Accreditation Board for Testing and Calibration Laboratories

Confirmmittee Equitee

Fig. 16.1: Symbols of importance

FURTHER READINGS

1. Koneman' Color Atlas and Textbook of Diagnostic Microbiology, 7th edition, Lippincott William and Wilkins.

2. Mackie and McCartney's, Practical Medical Microbiology, 14th edition, Churchill Livingstone, Elsevier

3. Bailey and Scott's Diagnostic Microbiology, 13th edition, Elsevier Health – US.

4. Mandell, Douglas and Bennett's Principles and Practice of Infectious Diseases, 7th edition, Elsevier Health – US.

5. Topley and Wilson's Microbiology and Microbial infection, 10th edition, Wiley-Blackwell.

Index